AF737993

ADVANCED TOPICS IN BIOMINERALIZATION

Edited by **Jong Seto**

Advanced Topics in Biomineralization
Edited by Jong Seto

Published by InTech

Janeza Trdine 9, 51000 Rijeka, Croatia

Copyright © 2012 InTech

All chapters are distributed under the Creative Commons Attribution 3.0 license, which allows users to download, copy and build upon published articles even for commercial purposes, as long as the author and publisher are properly credited, which ensures maximum dissemination and a wider impact of our publications. After this work has been published by InTech, authors have the right to republish it, in whole or part, in any publication of which they are the author, and to make other personal use of the work. Any republication, referencing or personal use of the work must explicitly identify the original source.

As for readers, this license allows users to download, copy and build upon published chapters even for commercial purposes, as long as the author and publisher are properly credited, which ensures maximum dissemination and a wider impact of our publications.

Notice

Statements and opinions expressed in the chapters are these of the individual contributors and not necessarily those of the editors or publisher. No responsibility is accepted for the accuracy of information contained in the published chapters. The publisher assumes no responsibility for any damage or injury to persons or property arising out of the use of any materials, instructions, methods or ideas contained in the book.

Publishing Process Manager Romana Vukelic

Technical Editor Teodora Smiljanic

Cover Designer InTech Design Team

Additional hard copies can be obtained from orders@intechopen.com

Advanced Topics in Biomineralization, Edited by Jong Seto
p. cm.
ISBN 978-953-51-0045-4

We are IntechOpen, the world's leading publisher of Open Access books

Built by scientists, for scientists

4,100+
Open access books available

116,000+
International authors and editors

120M+
Downloads

151
Countries delivered to

Our authors are among the
Top 1%
most cited scientists

12.2%
Contributors from top 500 universities

CLARIVATE ANALYTICS BOOK CITATION INDEX INDEXED

WEB OF SCIENCE™

Selection of our books indexed in the Book Citation Index in Web of Science™ Core Collection (BKCI)

Interested in publishing with us?
Contact book.department@intechopen.com

Numbers displayed above are based on latest data collected.
For more information visit www.intechopen.com

For EU product safety concerns:
IN TECH d.o.o., Prolaz Marije Krucifikse Kozulić 3, 51000 Rijeka, Croatia,
info@intechopen.com or visit our website at intechopen.com.

Meet the editor

Dr Jong Seto, from Los Angeles, California, USA, studied mineralization processes in echinoderms in larval model systems at the University of California, Berkeley and completed his Ph.D. in bone architecture and mechanics at the Max-Planck-Institute of Colloids and Interfaces in 2010. His current research interests include development of in situ measurements of organic-inorganic interactions in calcium carbonate and phosphate mineralization systems as well as structure-function relationships in biological materials. He is currently a research fellow working with Prof. Helmut Cölfen at the Universität Konstanz and enjoys cycling, hiking, as well as skiing outdoors in the nearby Alps.

Contents

Preface

Emerging from investigations of bone, shell, and tooth formation, the field of biomineralization has rapidly grown in the past decade—utilizing novel method and device developments that have resulted from advances in nano -science and -technology. These tools have enabled direct measurement, manipulation, and visualization of processes at or near the molecular level. Not only have the methods improved, but biomineralization has become multidisciplinary—relying on the active cooperation of molecular biologists, physical chemists, as well as materials scientists to approach questions from other perspectives. Active centers in biomineralization research are found throughout Germany, Israel, Japan, United Kingdom as well as the United States.

Biomineralization has become not only an interdisciplinary matter but also an international one, which benefits tremendously from the cooperative as well as coordination of research efforts stretching around the world. Pooling together the expertise at these centers and disciplines to focus on key issues in biomineralization, our understanding of the formation, regulation of properties and application of biominerailized materials have dramatically improved.

By examining biomineralized materials at smaller length scales, we can observe interfaces where organic and inorganic interactions occur. At these length scales, molecules have been found that can inhibit mineralization. Another category of molecules that can nucleate mineral are more elusive, but are believed to exist as well. The interactions of the organic with the developing mineral—in the form of ions and clusters of ions—have led to a better understanding of the influence of additives involved in mineralization. Several groups have also found the existence of amorphous phases and their significance in directing the formation of specific crystalline phases.

If we fully understand these principles at the atomic and molecular levels, bottom-up construction schemes can be utilized to make materials that are tailored with specific properties. Some groups have even gone into exploring these schemes for building materials for use in construction of large structures like bridges and buildings. In the next decade, we hope that the biomineralization community will continue to grow and develop, incorporating novel examples of Nature's biomineralization toolkit to create functional materials that will create a more clean, safe, and livable society.

Jong Seto
Konstanz
Germany

Part 1

Biomineralizing Schemes and Strategies

1

Intrinsically Disordered Proteins in Biomineralization

Magdalena Wojtas, Piotr Dobryszycki and Andrzej Ożyhar
Wroclaw University of Technology
Poland

1. Introduction

Intrinsically disordered proteins (IDPs) have the potential to play a unique role in the study of proteins and the relationships between structure and function. Intrinsic disorder affects chemical and cellular events such as cell signaling, macromolecular self-assembly, protein removal and crystal nucleation and growth. This chapter explores the structural principles by which IDPs act and reveals the prevalence of IDPs in the field of biomineralization. It has been demonstrated that proteins involved in biomineralization are frequently very extended and disordered. Moreover, the disordered structure is integral to how these proteins fulfill their functions. We have focused on the analysis of polypeptide folding, the role of post-translational modifications, predictions of the structural disorder and the degree of disorder in secondary structures. Computational and biophysical strategies to analyze the secondary structures and evaluate the degree and nature of "disorder" in proteins are described. Biomineralization is the result of the orchestration of a series of protein-protein, protein-mineral and protein-cell interactions. Identifying unfolded functional domains in cell signaling may have a great impact in the study of tissue regeration and biomineral formation. IDPs are typically organic components of biominerals. It is believed that they could act as a regulatory coordinator for specific interactions of many proteins, and thus many physiological processes such as formation of dentin and bone, the formation of sea urchin and crusteacean exoskeletons. Here, we review what is currently understood about the molecular basis of biomineral formation. This includes protein interaction with metal ions, post-translational modifications, interactions with other proteins, or other factors that induce the formation of crystal shape and size along with the proper polymorph selection in relation to the role of IDPs.

2. Intrinsically disordered proteins

The history of IDPs goes back to the 1960s, with Linus Pauling's observation of the existence of regions in proteins with a disordered structure (Pauling & Delbruck, M., 1940). However, only a small group of researchers like Dunker, Uversky, Wright, Dyson, Tompa and others during the next forty years demonstrated that it was possible to depart from the paradigm that a protein's function is closely affiliated with its structure (Dyson & P.E. Wright, 2005; Tompa, 2011; Uversky & Dunker, 2010; P.E. Wright & Dyson, 1999). Currently, it is believed that 20-50% of eukaryotic proteins contain at least one fragment belonging to the class of IDPs (Babu et al., 2011; Dunker et al., 2000). It is well known that globular proteins decrease their activity in a denatured state when a solution is subjected to high temperatures or chemical denaturants. The "structure-function" paradigm was not contested for many years until experimental data began to show that there was no stable three-dimensional structure for some protein fragments that had been attributed to particular functions. These proteins are in whole or in part, in contrast to globular proteins, heterogeneous ensembles of flexible molecules, unorganized and without a defined three-dimensional structure. According to these properties, the proteins are referred to as natively unfolded, intrinsically unfolded (IUP), intrinsically unstructured or intrinsically disordered (Dunker et al., 2005; Dyson & Wright, 2005; Tompa, 2005; Uversky, 2002). It has been previously shown that structural disorder is characteristic of proteins involved in important biological processes such as signal transmission, regulation of cell cycles, regulation of gene expression, activity of chaperone proteins, neoplastic processes, and biomineral formation (Dyson, 2011; Tompa, 2011; Uversky, 2010).

These processes require a series of dynamic macromolecular interactions and IDPs seem to be specially created for their functions. An IDP's meta-stable conformation allows it to bind to its protein partners as well as interact with high specificity and relatively low affinity. Furthermore, there is some experimental evidence showing that IDPs may interact with multiple partners, changing or adjusting the structures and functions of their partners (Tompa, 2005). Comparative analyses of the amino acid sequences of all currently known IDPs have shown common features. These proteins are characterized by amino acid compositions enriched with residues like A, R, G, Q, S, P, E and K that promote a disordered structure, with the small participation of other residues like W, C, F, I, Y, V, L and N, which simultaneously promote an ordered structure (Dunker et al., 2001). IDPs can be classified into five groups based on their relative functions: entropic chains, effectors, assemblers, scavengers, and display sites. Entropic chains act as flexible linkers between the globular domains of multidomain proteins. Effectors bind and modify the activity of a partner. Assemblers are able to simultaneously bind several ligands as multimolecular assemblies. Scavengers store or neutralize small ligands. Finally, display sites promote specific interactions within the active sites of enzymes that facilitate post-translational modifications (Tompa, 2002).

2.1 Methods for analyzing IDP structure

Based on the amino acid composition of IDPs, a number of algorithms have been proposed that predict regions containing a disordered structure. The most frequently used are PONDR (Romero et al., 2001), DISOPRED 2 (Ward et al., 2004a, 2004b), IUPred (Dosztanyi et al., 2005), GLOBPLOT 2 (Linding et al., 2004) and FoldIndex (Uversky et al., 2000). More algorithms can be found in the DisProt database (Sickmeier et al., 2007). These algorithms operate on the principle of a neuronal network "trained" using amino acid sequences belonging to experimentally confirmed IDPs. It has been observed that sequences that have low complexity or are abundantly charged and/or freuquently post-translationally modified (e.g. phosphorylated) usually adopt a stretched, unordered conformation (Romero et al., 2001). IDPs are often characterized by charge-hydropathy plots. Based on the normalized net charge and mean hydrophobicity, proteins can be categorized into either globular folded proteins or IDPs. IDPs are specifically localized within a unique region of charge-hydrophobicity proteins (Uversky et al., 2000). Figure 1A shows a charge-hydropathy plot for experimentally confirmed IDPs and globular proteins, while Figure 1B presents a charge-hydropathy plot for proteins involved in biomineralization.

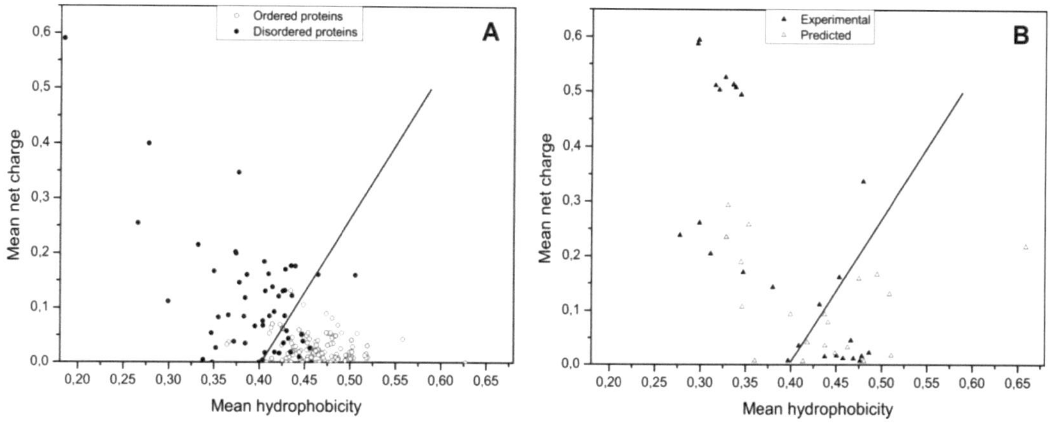

Fig. 1. Charge-hydrophobicity plots. The solid line represents the contractual boundary between disordered and ordered proteins. (A) Comparison of charge-hydrophobicity for IDPs (black dots) and globular proteins (white dots). (B) Proteins involved in biomineralization represented on the charge-hydrophobicity plot. Black triangles show experimentally confirmed IDPs, while white triangles represent proteins whose secondary structure has not been studied.

Even this simple analysis indicates that many proteins involved in biomineralization could be IDPs. However, computer predictions themselves can not be relied on as the sole evidence of the absence of structural order in a protein. Such evidence only invites further experimental study, which could include X-ray diffraction analysis (Dunker & Obradovic, 2001), multidimensional nuclear magnetic resonance spectroscopy (NMR) (Bai et al., 2001), circular dichroism spectroscopy (CD) (Tompa, 2002), differential scanning microcalorimetry (DSC) (Mendoza et al., 2003), X-ray scattering at small angle (SAXS) (Millett et al., 2002).

3. IDPs found in calcium phosphate related mineralization

In bone and dentin, collagen acts as a structural matrix whereas hydroxyapatite (HA) nucleation is regulated by the acidic phosphoproteins (Chen et al., 1992; Glimcher, 1989). These non-collagenous proteins (NCPs) play crucial roles in the organization of the collagen matrix and in the modulation of HA crystal formation (Ganss et al., 1999). NCPs are often classified as IDPs. Some examples of IDPs engaged in HA formation are presented below.

3.1 SIBLINGs

Small integrin-binding ligand, N-linked glycoproteins (SIBLINGs) with NCPs are involved in the mineralization of bone and dentin (George & Veis, 2008; Qin et al., 2004). Within the family of human SIBLINGs there is limited sequence similarity; however, they share common features, such as: (i) similar gene organization and chromosome localization, (ii) RGD (arginine-glycine-aspartate) motifs mediating cell attachment/signaling *via* their interactions with cell-surface integrins, (iii) extensive post-translational modifications like phosphorylation and glycosylation, (iv) abundance of acidic residues, (v) calcium ions and collagen binding ability (George & Veis, 2008; Qin et al., 2004), (vi) intrinsically disordered molecular character (Tompa, 2002). The SIBLINGs family includes osteopontin (OPN), bone sialoprotein (BSP), dentin matrix protein 1 (DMP1), matrix extracellular phosphoglycoprotein (MEPE) and dentin sialophosphoprotein (DSPP). DSPP gives rise to two mature products, dentin phosphoprotein (also called phosphophoryn) (DPP) and dentin sialoprotein (DSP) (George & Veis, 2008; Qin et al., 2004).

3.1.1 DMP1

The highly acidic protein (D and E constitute 29% of all residues) DMP1 acts as a nucleator for HA deposition *in vitro* (He et al., 2003a). The disordered character of DMP1 has been shown using several methods (Tab. 1). CD and FTIR measurements have shown that DMP1 has a random structure in solution, however upon calcium ions binding DMP1 undergoes a slight conformational change to a more ordered structure. SAXS and DLS confirmed a calcium-induced disorder-to-order transition in DMP1 leading to oligomerization (Gericke et al., 2010; He et al., 2003b). Moreover, it has been shown that the DMP1 molecule assumes an elongated shape (He et al., 2005a). Further studies have revealed that two specific acidic clusters (ESQES and QESQSEQDS) in DMP1 are responsible for the calcium-induced oligomerization and *in vitro* nucleation of apatite crystals (He et al., 2003b). This calcium-induced conformational change of DMP1 could be the structural basis for biocomposite self-assembly (He et al., 2003b).

The lack of a rigid structure enables DMP1 to serve multiple functions, not only in biomineralization, but also in osteoblast differentiation and maturation (Narayanan et al., 2003). Nonphosphorylated DMP1 is localized in the nucleus where it acts as a transcriptional component for the activation of matrix genes involved in mineralized tissue formation (Narayanan et al., 2003). It binds the DSPP gene promoter and activates DSPP gene expression. Calcium ions released from intracellular stores bind DMP1 and induce in DMP1 a conformational change and phosphorylation by casein kinase 2 (CK2). Finally, the phosphorylated protein is exported to the extracellular matrix, where it acts as a nucleator of hydroxyapatite (Narayanan et al., 2003). The DNA binding domain is localized within the C-terminal region of DMP1 (Narayanan et al., 2006). Extracellular DMP1 also has the ability to strongly bind the H factors, integrin $\alpha v\beta 3$ and CD44 (Jain et al., 2002), and it is specifically involved in signaling via extracellular matrix-cell surface interaction (Wu et al., 2011).

Organism	Protein	pI	Methods	Reference
Mammals	DMP1	4.0	CD, DLS, FTIR, SAXS	(Gericke et al., 2010; He et al., 2003a, 2003b)
	DPP	2.8	CD, NMR, SAXS	(Cross et al., 2005; Evans et al., 1994; Fujisawa & Kuboki, 1998; George & Hao, 2005; He et al., 2005b; Lee et al., 1977)
	BSP	4.1	CD, NMR, SAXS	(Fisher et al., 2001; Tye et al., 2003, 2005; Wuttke et al., 2001)
	OPN	4.4	CD, NMR, FTIR	(Fisher et al., 2001; Gorski et al., 1995)
	amelogenin	6.6	CD, NMR	(Buchko et al., 2010; Delak et al., 2009b; Ndao et al., 2011;Shaw et al., 2008)
	statherin	8.0	CD, NMR	(Long et al., 2001; Naganagowda et al., 1998; Raj et al., 1992)
	lithostathine	5.7	CD	(Gerbaud et al., 2000)
Haliotis rufescens	AP7	5.2	CD, NMR	(Kim et al., 2004, 2006a; Michenfelder et al., 2003; Wustman et al., 2004)
	AP24	5.3	CD, NMR	(Michenfelder et al., 2003; Wustman et al., 2004)
	Lustrin A	8.1	NMR	(Wustman et al., 2003; Zhang et al., 2002)
Picntada fucata	n16	7.5	CD, NMR	(Amos et al., 2011; Collino & Evans, 2008; Kim et al., 2004, 2006b)
	ACCBP	4.7	CD	(Amos et al., 2009)
	PFMG1	7.9	CD	(Liu et al., 2007)
Atrina rigita	Asprich	2.7-3.5	CD, NMR	(Collino et al., 2006; Delak et al., 2009a, 2008; Kim et al., 2008; Ndao et al., 2010)
Procambrus clarkii	CAP-1	3.9	CD	(Inoue et al., 2007)
Strongylocentrotus purpuratus	SM50	10.8	CD, NMR	(Xu & Evans, 1999; Zhang et al., 2000)
	PM27	8.1	CD, NMR	(Wustman et al., 2002)
Danio rerio	Stm	4.1	CD, gel filtration	(Kaplon et al., 2008, 2009)

Table 1. IDPs involved in biomineralization of calcium carbonate and phosphate for which a disordered structure has been confirmed experimentally.

3.1.2 DSPP

DSPP undergoes proteolytic cleavage to DPP and DSP. DPP is most abundant in dentin, but also present in bone (George & Hao, 2005; Lee et al., 1977). At least 75% of the DPP sequence (isolated from dentin) is composed of S and D residues and 85-90% of the S residues are phosphorylated (George & Veis, 2008; He et al., 2005b; Huq et al., 2000). In solution DPP isolated from dentin behaves as a fairly extended, random-chain molecule due to electrostatic repulsion (Table 1), as has been shown by CD studies (Lee et al., 1977). The presence of calcium ions reduces DPP solubility, which indicates aggregation of the protein (Lee et al., 1977). It has been demonstrated that nonphosphorylated DPP has lower calcium binding ability than the phosphorylated form and induces amorphous calcium phosphate formation, while the phosphorylated form promotes plate-like apatite crystals (He et al., 2005b). SAXS studies revealed a calcium-induced conformation change from an extended structure to a more compact one, but only in phosphorylated DPP. Nonphosphorylated DPP was disordered irrespective of the presence of calcium ions at various concentrations (George & Hao, 2005; He et al., 2005b). NMR spectroscopy also confirmed high mobility and flexibility of DPP in the absence of calcium ions, and decreased mobility in the presence of calcium ions (Cross et al., 2005; Evans et al., 1994). Solid-state NMR spectroscopy enabled investigation of the DPP structure when bonded to HA. The secondary structure of DPP bound to crystal was very extended and largely disordered. A majority of DPP residues interacted with crystal. (Fujisawa & Kuboki, 1998). This disordered molecular structure facilitates DPP's extension across the surface of a crystal and allows it to cover the surface with only a small number of molecules, all of which results in a highly inhibitory effect on crystal growth (Fujisawa & Kuboki, 1998).

While DPP structure has been extensively explored, DSP structural studies are still unavailable. However, bioinformatic predictions strongly suggest that DSP is also an IDP (Table 2).

			Overall percent disordered		
Organism	**Protein**	**pI**	**IUPred**	**DISOPRED2**	**PONDR**
Mammals	DSP	4.5	100	100	73
	MEPE	8.6	95	60	64
Picntada fucata	Aspein	1.5	100	100	100
	Prismalin-14	3.9	0	0	9
Patinopecten yessoensis	MSP-1	3.2	97	98	96
Pinna nobilis	Calprismin	4.9	12	0	74
Crassostrea nippona	MPP1	2.2	100	98	97
Nautilus macromphalus	Nautilin-63	9.2	66	4	41
Procambrus clarkii	GAMP	4.2	63	73	63
	CAP-2	4.2	18	2	45
	Casp-2	4.3	33	8	68
Orchestia cavimana	Orchestin	4.5	58	34	69
Cherax quadricarinatus	GAP 65	5.0	2	0	11
	GAP 10	5.5	4	0	37
Penaeus japonicus	Crustocalcin	3.9	45	37	81
Strongylocentrotus purpuratus	SM30	6.2	25	0	28
	SM32	8.3	44	44	47
	SM37	10.4	60	43	56
	phosphodontin	3.9	96	94	94
Oryzias latipes	Stm-like	3.8	100	98	94
Oncorhynchus mykiss	OMM-64	3.5	97	96	92
Gallus gallus	Ovocleidin-116	6.6	91	57	78

Table 2. Bioinformatic predictions of a disordered structure in proteins involved in biomineralization.

All three bioinformatic tools (IUPred, DISOPRED2, and PONDR) predict that DSP is largely or completely disordered. DSP, similarly to DPP, also has a low pI value, which causes electrostatic repulsion that leads to an extended structure. However, this presumption requires experimental confirmation.

3.1.3 BSP

BSP is a multifunctional protein involved in cell attachment, signaling, HA binding, HA nucleation and collagen binding (Ganss et al., 1999). BSP is relatively specific to skeletal tissues and is the most abundant protein in bone (Bianco et al., 1991). NMR, CD and SAXS studies of BSP showed a high level of disordered structure (Table. 1) (Fisher et al., 2001; Tye et al., 2003, 2005; Wuttke et al., 2001). Post-translationally, unmodified BSP has lower rates of migration during SDS-PAGE, as do other IDPs (Stubbs et al., 1997). BSP visualized by electron microscopy after rotary shadowing was an extended monomer possessing a globular structure, probably on the C-terminus of the protein (Wuttke et al., 2001). BSP treated with 6 M Gdm-HCl lost its residual α-helix and β-sheet structure. Structural studies indicated that BSP had IDP-like characteristics, but not a random structure (Wuttke et al., 2001). Calcium ions had a significant effect on BSP conformation, in contrast to DMP1 and DPP (Tye et al., 2003; Wuttke et al., 2001). The flexibility and plasticity of BSP may enable cell attachment to HA. Additionally, BSP has a strong affinity for HA (Fujisawa et al., 1997), while on the other hand, RGD sequences allow integrin binding (Fujisawa et al., 1997; Oldberg et al., 1988). Hence, the extension and flexibility of BSP may be advantageous to its function as a bridge for the cell attachment of HA (Tye et al., 2003). The intrinsic disorder of BSP seems to be important for interacting with type I collagen (Tye et al., 2005).

3.1.4 OPN

While DPP, BSP and DMP1 are relatively specific to bone and teeth, OPN is also present in the brain, kidney, smooth muscle, macrophages, inner ear and body fluids (milk, urine, bile) (George & Veis, 2008; Gericke et al., 2005; Kazanecki et al., 2007). Its ubiquitous expression pattern and variations in phosphorylation, glycosylation and sulphation suggest that OPN fulfills many different functions (George & Veis, 2008). Moreover, OPN controls the formation of calcium phosphate (Goldberg & Hunter, 1995), calcium carbonate (Chien et al., 2008), and calcium oxalate (Grohe et al., 2007) crystals. NMR, CD and FTIR studies demonstrated that OPN is an IDP in solution (Tab.1) (Fisher et al., 2001; Gorski et al., 1995). The addition of calcium ions had little effect on the CD spectrum of OPN, while increasing protein concentration led to a more organized secondary structure (Gorski et al., 1995). Hence, it was suggested that OPN conformation in solution may be different from the conformation of OPN adsorbed to HA, because the local concentration of OPN in the semi-solid matrix might be relatively high (Gorski et al., 1995). This theory is supported by the observation that antibodies do not recognize OPN bound to HA, while they are able to recognize OPN bound to plastic (Gorski et al., 1995). Moreover, the acidic peptide based on the OPN sequence, which was examined by RosettaSurface, was able to adsorb to a calcium oxalate crystal surface in multiple conformations (Chien et al., 2009).

OPN is a substrate of tissue transglutaminase (Prince et al., 1991). Investigations of polymeric OPN showed a 5-fold increase in OPN binding to collagen type I when it was a polymer than when it was a monomer (Kaartinen et al., 1999), which indicates that there is a conformational change in OPN upon cross-linking. CD measurements demonstrated that cross-linked OPN has a more organized structure than monomeric OPN (Kaartinen et al., 1999). It seems that OPN needs molecular crowding to assume a more ordered structure.

OPN found in non-mineralized tissues may or may not be phosphorylated, while bone OPN is always phosphorylated (George & Veis, 2008; Veis et al., 1997). Low phosphorylated OPN from bone (38%) is an effective HA mineralization inhibitor in contrast to highly phosphorylated OPN from milk (96%), which promotes HA formation and growth. The dephosphorylated form had no significant effect (Gericke et al., 2005). The degree to which OPN affected HA formation depended on the level of phosphorylation (Gericke et al., 2005). It has been suggested that the binding of OPN to HA alters OPN conformation, facilitating recruitment and activation of macrophages that remove pathologic HA deposits (Steitz et al., 2002). It was also shown that dephosphorylated OPN loses its ability to inhibit smooth muscle cell calcification (Jono et al., 2000).

3.2 Enamel proteins

Amelogenins are a protein family derived from a single gene by alternative splicing and controlled, post-secretory processing. They are abundant in tooth enamel, but amelogenin has also been identified in dentin, bone, cartilage and nonmineralizing tissues such as those of the brain, salivary glands and macrophages (Gruenbaum-Cohen et al., 2009; Lyngstadaas et al., 2009). Amelogenin is involved in biomineralization as well as cell signaling events (Gruenbaum-Cohen et al., 2009; Lyngstadaas et al., 2009; Veis, 2003). The primary sequence of amelogenin is highly conserved, especially the N-terminal Tyr-rich and C-terminal charged regions (Delgado et al., 2005). Determination of amelogenin's secondary structure has been hampered by the self-association of the protein (Li et al., 2006). Far UV CD and NMR spectra demonstrated that amelogenin in a monomeric form is largely disordered (Table 1). There is no well-defined, continuous region with an ordered structure, but some residual secondary structure was observed. In addition, amelogenin exists in at least two conformations (Buchko et al., 2010; Delak et al., 2009b; Ndao et al., 2011). Amelogenin can interact with other important proteins engaged in enamel formation (Bartlett et al., 2006). Amelogenin binds to CD63 and LAMP1 receptors, which mediate signal transduction events and endo-, pino-, and phagocytosis, respectively. Both partners interact with the same amelogenin motif, which is largely disordered and accessible to the external environment (Shapiro et al., 2007; Zou et al., 2007).

The large number of alternatively spliced variants, expression in different tissues and the intrinsically disordered character of ameloganin reflect the multifunctionality of the protein, which is probably a common feature among IDPs.

Ameloblastin is also involved in enamel biomineralization, interactions between the ameloblasts and enamel extracellular matrix, and regeneration of hard-tissue. No NMR or CD studies of ameloblastin are available; however, bioinformatic analysis and molecular modeling strongly suggest that the protein is an IDP (Vymetal et al., 2008).

3.3 Statherin

Statherin is an inhibitor of the nucleation and growth of HA in the supersaturated environment of saliva (Hay et al., 1984; Johnsson et al., 1991; Stayton et al., 2003). The N-terminal region of statherin contains highly acidic motifs (DSpSpEE, where Sp indicates a phosphoseryl residue), which are important for adsorption of statherin onto the surface of HA and the inhibition of HA growth (Raj et al., 1992). Statherin was reported as unstructured in solution using NMR techniques (Naganagowda et al., 1998). Far UV CD experiments showed that N-teminal regions of statherin had a strong tendency to adopt an α-helical structure (upon the addition of TFE), while middle and C-terminal portions of this protein were flexible and preferred to adopt an unordered conformation (Naganagowda et al., 1998; Raj et al., 1992). Solid-state NMR applied to examine the structure of statherin bound to HA (Tab. 1) demonstrated, that under biologically relevant conditions, 12 residues of the N-terminus of statherin were in a helical conformation and that they were strongly adsorbed to the HA surface. In contrast, middle and C-terminal regions weakly interacted with HA and were highly mobile. The mobility of statherin on HA surfaces could allow it to effectively block more nucleation sites than a very rigidly bonded protein (Long et al., 2001).

4. IDPs found in calcium carbonate related mineralization

Calcium carbonate biominerals are widespread in nature. They occur in vertebrates as well as invertebrates, where they fulfill various functions. Here, we present the characteristics of IDPs involved in calcium carbonate crystal formation.

4.1 Lithostathine

Pancreatic fluid is supersaturated with calcium and carbonate ions; however, stones are observed only in cases of patients with chronic calcifying pancreatitis (De Caro et al., 1988; Moore &Verine, 1987). Lithostathine, a pancreatic glycoprotein, inhibits the growth and nucleation of calcium carbonate crystals. The C-terminal region of lithostathine is homologous with C-type lectins (Patthy, 1988). The crystal structure of lithostathine has been determined. However, the 13 residues of the N-terminus, including the glycosylation site, are flexibile in the crystal (Bertrand et al., 1996). Far UV CD spectra of the N-terminal region are typical for random-coil structures (Table 1) (Gerbaud et al., 2000). Interestingly, the short N-terminal peptide of lithostathine is essential for the inhibition of nucleation and crystal growth. The glycosylated N-terminal peptide generated by limited trypsin hydrolysis inhibited crystal growth similarly to full-length lithostathine, while the C-terminal polypeptide was inactive. A synthetic N-terminal peptide, but not glycosylated was equally active (Bernard et al., 1992). Hence, it was postulated that backbone flexibility is essential for the inhibitory effect of the protein (Gerbaud et al., 2000).

4.2 Mollusk shell proteins

The mollusk shell is a commonly used model system for studying calcium carbonate biomineralization. Some mollusks have developed a bilayer of composite materials. Interestingly, the two layers, prismatic and nacreous, are composed of different forms of calcium carbonate polymorph, calcite and aragonite, respectively (Evans, 2008).

Several acidic proteins from *Haliotis rufescens* have been identified and characterized. AP7, AP8 and AP24 (aragonitic protein of molecular weight 7kDa, 8kDa and 24kDa, respectively) have been identified from the demineralized nacre protein fraction (Fu et al., 2005; Michenfelder et al., 2003). *In vitro* mineralization studies showed that AP7 and AP24 interacted with the step edges of growing calcite, which inhibited growth (Michenfelder et al., 2003). Analysis of the AP7 and AP24 sequence revealed that the N-terminus of both proteins (1-30 residues of the mature protein) is characterized by calcite binding domains. Both sequences are capable of affecting calcium carbonate crystal growth *in vitro* (Michenfelder et al., 2003). Far-UV CD studies of AP7 showed that the N-terminus of AP7 adopts a random-coil or extended conformation in solution (Tab 1.) (Michenfelder et al., 2003; Wustman et al., 2004), while the C-terminus is α-helical (Kim et al., 2006a). Interestingly, AP7's C-terminus end did not influence crystal growth (Kim et al., 2006a), while AP7's N-terminus was responsible for interactions with calcium ions and led to an inhibition of calcite growth *in vitro* (Kim et al., 2004; Michenfelder et al., 2003). It should be noted that full-length AP7 protein had the highest effect on calcite crystal growth (Kim et al., 2006a). This observation suggests that the flexibility and plasticity of IDPs might be crucial factors which inhibit crystal growth, although the well-defined, ordered secondary structure may also be necessary. It should be noted that high concentrations of calcium ions led to the precipitation of AP7's N-terminus, which may indicate calcium-dependent oligomerization (Kim et al., 2004). Finally, full-length AP7 undergoes a conformational change to an α-helical structure as a function of TFE concentration (Amos & Evans, 2009).

Another protein from *Haliotis rufescens*, Lustrin A, is a component of the intercrystalline organic matrix lying between layers of aragonite tablets. The postulated role of Lustrin A is to enhance nacre layers' resistance to fracture (Shen et al., 1997). NMR studies showed that the N-terminal end of Lustrin A contains loop regions, which behave as entropic springs that endow it with resilience and flexibility (Table 1). This sequence is believed to be one of several putative elastic motifs within Lustrin A (Zhang et al., 2002). Although the sequence adopts a relatively defined secondary structure, the plasticity of Lustrin A's loop regions seems to be crucial to its postulated role in resisting fracture. Further studies of model peptides based on Lustrin A sequences showed two domains (RKSY and D4) existing largely in conformationally labile random-coil and extended states (Wustman et al., 2003). Possibly, these features permit side chain accessibility to exposed calcium carbonate crystal surfaces (Wustman et al., 2003). In addition, its D4 domain, which as the name implies contains 4 D residues, inhibits nucleation of calcite crystals *in vitro* (Wustman et al., 2003). The inhibitory effect strongly supports the theory that IDPs control biomineralization processes.

Another mollusk protein, n16, has been isolated from the nacre layer of *Picntada fucata* (Samata et al., 1999). The analysis of n16's primary sequence revealed the requisite amino acid residues associated with a calcium carbonate modification domain within the C- and N-terminae of the mature protein (n16-C and n16-N, respectively) (Kim et al., 2004). Both sequences modify the morphology of calcium carbonate crystals and induce calcite growth *in vitro* (Kim et al., 2004). Far-UV CD and NMR studies demonstrated that n16 exists as a random-coil molecule; however, increasing the concentration of the peptide led to a conformational equilibrium between random-coil and β-sheet conformers (Tab. 1) (Kim et al., 2004). Interestingly, in the presence of TFE, n16-N adopted a β-sheet conformation, unlike AP7 and AP24, which adopted an α-helical conformation. Thus, AP7, AP24 and n16N are conformationally labile, but exhibit different folding propensities (Collino & Evans, 2008). Moreover, random scrambling of n16-N abolishes its concentration-dependent conformational rearrangement capability, resulting in a decrease of its ability to affect crystal growth (Kim et al., 2006b). It was also shown that n16-N simultaneously interacted with β-chitin and the nucleating mineral phase, finally leading to aragonite formation *in vitro* (Keene et al., 2010). Also noteworthy, n16-N alone induced calcite formation (Kim et al., 2004). It has been suggested that the binding of n16-N to β-chitin could trigger disorder-to-order transitions. In summary, the intrinsically disordered structure of n16-N facilitates interactions with its partners (like for example β-chitin) and disordered-to-ordered transition seems to be crucial for self-assembly (Amos et al., 2011; Collino & Evans, 2008; Keene et al., 2010).

ACCBP (amorphous calcium carbonate binding protein) was found in the extrapallial fluid of *Picntada fucata* (Ma et al., 2007), while PFMG1 (*Picntada fucata* mantle gene protein 1) controls calcium carbonate nucleation and is believed to assist in the formation of pearl nacre (Liu et al., 2007). Based on previously obtained results (Amos et al., 2009; Collino et al., 2006; Keene et al., 2010; Kim et al., 2004; Wustman et al., 2004) and bioinformatic predictions, the C- and N- terminal sequences of both proteins (PFMG1-C, PFMF1-N and ACCN) have been studied to find mineralization activity and conformational disorder (Amos et al., 2009; Liu et al., 2007). Far UV CD spectroscopy confirmed that PFMG1-C, PFMF1-N and ACCN possess a combination of a random-coil conformation and other secondary structures (Table 1). Moreover, the addition of TFE led to disorder-to-order transitions in all peptides (Amos et al., 2009; Liu et al., 2007).

Aspein is another extremely acidic (60.4 % of D) protein from *Picntada fucata* (Tsukamoto et al., 2004). The extraordinary acidic character of Aspein and the the fact that SDS-PAGE overestimated its molecular mass (Takeuchi et al., 2008; Tsukamoto et al., 2004) suggested that Aspein is also an IDP. Bioinformatic tools predicted that Aspein is completely disordered (Table 2). Prismalin-14 from the prismatic layer of *Picntada fucata* also contains putative intrinsically disordered regions. Firstly, G/Y-rich regions might form glycine loop motifs, which may contribute to the elasticity of the molecule (Suzuki et al., 2004). This motif is responsible for interaction with chitin (Suzuki & Nagasawa, 2007). Secondly, D-rich regions within the N- and C- terminae of Prismalin-14, which are responsible for inhibiting calcium carbonate precipitation *in vitro* (Suzuki & Nagasawa, 2007), also might be intrinsically disordered. However, only PONDR predicted short intrinsically disordered regions in Prismalin-14. IUPred and DISOPRED2 didn't predict any disordered structure (Table 2).

Ten acidic Asprich proteins (51-61% of acidic residues) have been identified from *Atrina rigita* (Gotliv et al., 2005). The Asprich family might be the product of alternative RNA splicing of the same gene and are arbitrarily divided into six domains: (i) N-terminal signal peptide, (ii) basic domain, (iii) the acidic1 domain, identical in all the Asprich proteins, (iv) the variable acidic domain, which differs among Asprich proteins, (v) conserved DEAD repeat domain, (vi) and conserved acidic2 domain (Gotliv et al., 2005). All three conserved acidic domains (acidic1, DAED and acidic2) are intrinsically disordered and they control calcium carbonate crystal growth (Collino et al., 2006; Delak et al., 2009a; Kim et al., 2008). Far UV CD studies of acidic1, DAED and acidic2 domains demonstrated that these peptides exist in a random-coil conformation in equilibrium with other secondary structures. Interestingly, a significant conformational change was not observed at lower pH and in the presence of calcium ions, which suggests that structural lability is a result of more than just electrostatic repulsion (Collino et al., 2006; Delak et al., 2009a; Kim et al., 2008). NMR spectroscopy showed that calcium ions induced local perturbations in conformations of acidic1, DAED and acidic2, but the structure was still very flexible (Table 1) (Delak et al., 2009a). Asprich proteins are able to adopt a more ordered structure in the presence of TFE, indicating that they might undergo a disorder-to-order transition (Ndao et al., 2010).

Several other acidic proteins have been identified from mollusk shells, for instance: MSP-1 (molluskan shell protein 1) from *Patinopecten yessoensis* (Sarashina& Endo, 2001), Caspartin and Calprismin from *Pinna nobilis* (Marin et al., 2005), moluskan phosphorylated protein 1 (MPP1) from *Crassostrea nippona* (Samata et al., 2008), P95 from *Unio pictorum* (Marie et al., 2008), and Nautilin-63 from *Nautilus macromphalus* (Marie et al., 2011). No secondary structure analyses of these proteins have been published, hence it is unknown if they are IDPs. However, MSP-1 and MPP1 are likely to be largely disordered, while Nautilin-63 probably consists of a combination of disordered and ordered regions (Table 2).

4.3 Crusteacean associated IDPs

Most crustaceans possess a calcified exoskeleton that is periodically molted to permit growth. Hence, their calcium metabolism is tightly linked to seasonal molting cycles. Aquatic crustaceans take calcium from the water, whereas terrestrial species have developed several storage strategies (Luquet et al., 1996; Testeniere et al., 2002). During the premolting period, calcium is stored in the hemolymph, the hepatopancreas, sterna plates, gastroliths or posterior caeca, depending on the species (Luquet et al., 1996; Meyran et al., 1984). All calcified deposits correspond to biomineralized structures (Testeniere et al., 2002). After ecdysis, stored calcium is quickly reabsorbed and translocated to harden the new exoskeleton (Ishii et al., 1996). The exoskeleton of crustaceans consists of calcium carbonate and chitin-protein microfibrillar frameworks.

Calcification-associated peptide 1 (CAP-1) is an acidic protein isolated from the exoskeleton of *Procambrus clarkii* (Inoue et al., 2001). The C-terminal region is especially acidic and possesses a phosphoseryl residue, which is essential for the inhibition of calcium carbonate precipitation and the calcium-binding activities of CAP-1 (Inoue et al., 2003, 2007). Far UV CD studies demonstrated that CAP-1 has a random-coil conformation in equilibrium with a residual β-sheet (Table 1). The N-terminal region contributes to maintaining the overall conformation of CAP-1 (Inoue et al., 2007). It has been suggested that the N-terminal fragment of CAP-1 inhibits crystal growth less efficiently, because it possesses a more ordered structure than the C-terminal fragment (Inoue et al., 2007). This hypothesis is supported by the observation that CAP-2, which has a sequence similar to CAP-1, but shorter by 17 residues at the C-terminus, is a less active inhibitor of calcium carbonate precipitation (Inoue et al., 2004, 2007).

There are more proteins identified from the calcium storage organs of exoskeleton in crustaceans like CAP-2 (Inoue et al., 2004), Casp-2 (calcification-associated soluble protein-2) (Inoue et al., 2008) and GAMP (gastrolith matrix protein) (Ishii et al., 1996) from *Procambarus clarkii*, orchertin from *Orchestia cavimana* (Hecker et al., 2004; Luquet et al., 1996), gastrolith protein 65 (GAP 65) (Shechter et al., 2008) and gastrolith protein 10 (GAP 10) (Glazer et al., 2010) from *Cherax quadricarinatus*, and crustocalcin (CCN) from *Penaeus japonicus* (Endo et al., 2004). Most of them are acidic; however, no secondary structure analysis has been performed. Bioinformatic predictions indicated that all of these proteins might possess intrinsically disordered regions (Table 2).

4.4 Sea urchin larval spicule matrix protein IDPs

Sea urchins have an elaborate, calcified skeleton, which is a valuable model system for analyzing molecular regulation of biomineralization (Wilt & Ettensohn, 2007; 2008). The skeleton is a network of calcified spicules, consisting of calcium and magnesium carbonate (in a 19:1 ratio), a surrounding extracellular matrix and small amounts of occluded matrix proteins (Wilt & Ettensohn, 2007; 2008). It has been estimated that about 40-50 different proteins are associated with the spicule (Killian & Wilt, 1996). SM30, SM32, PM27, SM37, and SM50 have been identified in *Strongylocentrotus purpuratus* (Benson et al., 1987; George et al., 1991; Harkey et al., 1995; Katoh-Fukui et al., 1991; Lee et al., 1999). SM32, PM27, SM37, and SM50 are basic, non-glycosylated, closely related proteins. They contain SP, C-type lectin domains and a variable number of proline-rich repeats (Killian & Wilt, 2008; Wilt, 1999; Wilt & Ettensohn, 2007; 2008).

SM50 is involved in the depositing initial calcite crystals as well as the elongation of the spicule elements, and it probably interacts with growing calcium carbonate crystals (Killian & Wilt, 2008). The C-terminal domain of SM50 contains two motifs exhibiting traits that are characteristic for IDPs. Firstly, GVGGR and GMGGQ repeats are present, which are sequence homologs of elastin and spider silk protein repeats. NMR studies of the C-terminal fragment showed that it exists as a β-sheet with an extended structure (Tab. 1). This structure might play an important role in matrix assembly, protein stability, molecular elasticity and protein-crystal recognition within the spicule's mineralized matrix (Xu & Evans, 1999). Secondly, PNNP repeat motifs are believed to play an important role in mineralization. P-rich sequences often exhibit the extended, flexible structure responsible for protein-protein and/or protein-crystal recognition, while N residues represent putative sites for side chain hydrogen bonding (Zhang et al., 2000). PNNP repeats adopt an extended, twisted conformation consisting of a mixture of turn-like and coil-like regions, most likely interconverting with the coiled-coil state (Zhang et al., 2000). Several possible roles for PNNP have been suggested: (i) as a mineral recognition domain, (ii) as a self-assembly domain, (iii) as a molecular spacer (Zhang et al., 2000).

PM27 contains PGMG repeated motifs, which have been examined by far UV CD, and NMR spectroscopy (Table 1). It has also been shown that this motif exhibits a random-coil or loop-like structure characterized by the ongoing oscillations between different conformational states (Wustman et al., 2002).

Bioinformatic tools also predicted that SM30, SM32, and SM37 might be partially disordered (Tab. 2). SM30, the most abundant spicule matrix protein, an acidic glycoprotein, also contains signal peptide and C-type lectin domains (Killian & Wilt, 1996; Wilt, 1999). It is coded by six genes, which are variously expressed in different mineralized structures (Illies et al., 2002; Killian et al., 2010; Livingston et al., 2006). Another protein isolated from *Srongylocentrotus purpuratus* is phosphodontin. It is the major acidic phosphoprotein of the tooth matrix (Mann et al., 2010). Phosphodontin is probably an IDP. All three predictors indicated that phosphodontin is disordered (Table 2).

4.5 Otolith related IDP

Otoliths are mineral deposits in the inner ear of fish, responsible for sensing gravity and detecting linear acceleration. They are mainly composed of calcium carbonate, but also contain a small fraction of organic molecules (1-5 % w/w). The organic matrix acts as a template for depositing crystals, promoting crystal growth in certain directions, and inhibiting crystal growth in undesired directions (Allemand et al., 2007; 2008; Ross & Pote, 1984).

Starmaker (Stm) and otolith matrix macromolecule-64 (OMM-64) are acidic proteins from *Danio rerio* and *Oncorhynchus mykiss* otoliths, respectively (Sollner et al., 2003; Tohse et al., 2008). Both proteins have similar gene organization and 25% sequence identity, and because of this, it has been postulated that they are orthologs (Sollner et al., 2003; Tohse et al., 2008). It should also be noted that DSPP is the close human functional analog of Stm (Sollner et al., 2003). Stm is responsible for the size, shape and polymorph selection in otoliths (Sollner et al., 2003). It has been demonstrated by several methods that Stm is an IDP (Tab. 1) (Kaplon et al., 2008, 2009). Far UV CD spectrum of Stm yielded the signatures typical for a disordered protein. The extended conformation of Stm results in the large hydrodynamic radius of the molecule, as was shown by gel filtration and ultracentrifugation. Interestingly, calcium ions led to a lower hydrodynamic radius, probably by decreasing electrostatic repulsion. Also, solvent-induced denaturation of Stm indicated the lack of an ordered structure in the molecule (Kaplon et al., 2008, 2009). Recently, another homolog of *Starmaker* has been identified: *Starmaker-like* from *Oryzias latipes*. The deduced amino acid sequence of Starmaker-like is also rich in acidic residues (Bajoghli et al., 2009). No secondary structure analysis of Stm-like, or OMM-64, have been performed; however, bioinformatic predictions strongly suggest that these proteins might be completely disordered (Table 2).

4.6 Egg shell related IDPs

An eggshell is a highly ordered natural, acellular composite bioceramic. It is one of the fastest and also one of the most structurally complicated biomineralizing systems. To produce an eggshell, 5 grams of calcium carbonate must be crystallized within 20 h (Arias et al., 2007; 2008). OPN, which is one of the eggshell proteins, was described above and is characterized as an IDP (3.1.4) (Chien et al., 2008; Panheleux et al., 1999). There are some proteins that are unique to eggshells. Ovocleidin-116 has been isolated and cloned from chickens (Hincke et al., 1999; Mann et al., 2002). Recent studies demonstrated that Ovocleidin-116 is an ortholog of mammalian MEPE, which belongs to the SIBLINGs family (Bardet et al., 2010). Although, structural studies of MEPE have not been available, it's possible that both proteins are IDPs, as was predicted by bioinformatic tools (Tab. 2). Several proteins identified from different avian species are C-type lectin-like proteins (Arias et al., 2007; 2008): ovocleidin-17 (chicken) (Hincke et al., 1995), ansocalcin (goose) (Lakshminarayanan et al., 2003), stuthiocalcin-1 and struthiocalcin-2 (ostrich) (Mann & Siedler, 2004), rheacalcin-1 and rheacalcin-2 (rhea) (Mann, 2004), dromaiocalcin-1 and dromaiocalcin-2 (emu) (Mann & Siedler, 2006). Structural studies of ovocleidin-17 and ansocalcin revealed that their three-dimensional structure is very similar to the structure of lithostathine (Bertrand et al., 1996; Lakshminarayanan et al., 2005; Reyes-Grajeda et al., 2004); however, the three proteins had different effects on calcium carbonate crystals. Lithostathine was an inhibitor, ansocalcin caused the formation of crystal aggregates, and ovocleidin-17 led to the crystals twinning (but not to an extensive aggregation of crystals like with ansocalcin) (Geider et al., 1996; Lakshminarayanan et al., 2005). It should be noted that N-terminal regions of ansocalcin and ovocleidin-17 are well folded, while lithostathine's N-terminus is disordered. Possibly lithostathine, ansocalcin and ovocleidin-17 influence crystal formation in different ways because of the distinct structure of each of their N-terminae. The disordered structure of lithostathine's N-terminus might be directly responsible for its inhibitory effect on crystal formation. Hence, based on the evidence presented in this review, it can be postulated that IDPs are crucial to the inhibition of crystal growth.

5. The innate specialization of IDPs in biomineralization

All of the above-mentioned examples strongly suggest that it is no accident that so many proteins involved in biomineralization exhibit an IDP-like character or are simply IDPs. Conformational instability is a common feature of proteins that bind to inorganic solids (Delak et al., 2009a; Kim et al., 2006a; Michenfelder et al., 2003). Intrinsic disorder provides benefits to these proteins and enables them to fulfill their functions.

Many IDPs involved in biomineralization have an unusual amino acid composition, for instance, they are rich in acidic residues. This unusual composition leads to strong electrostatic repulsion and results in the extended conformation of the IDP molecule.

The extended conformation of IDPs provides a much larger binding surface in comparison to the compact conformation of globular proteins. It is highly advantageous, especially when interacting with crystalline surfaces. Many proteins act as inhibitors of crystal growth by interacting with the crystal lattice and blocking nucleation sites. DPP's larger binding surface enables it to block more nucleation sites, hence one extended molecule can do the work of several molecules with a globular conformation (Fujisawa & Kuboki, 1998). Moreover, CAP-1, statherin and lithostathin structural studies clearly demonstrated that the highly flexible regions of these proteins were directly responsible for the inhibitory effects they had on crystal growth (Gerbaud et al., 2000; Inoue et al., 2007; Long et al., 2001).

Secondly, their extended conformation combined with their acidic character facilitates protein interactions with counter ions (Uversky, 2009). DPP and DMP1 bind relatively large amounts of calcium ions compared to calmodulin, which has a well-defined three-dimensional structure (He et al., 2003b, 2005b). Counter ion-binding seems to be crucial to weakening electrostatic repulsion and permitting the formation of β-sheets that lead to disorder-to-order transitions (He et al., 2005a). Asprich and AP7 proteins can bind calcium ions as well as other ions; however, this does not lead to significant global conformational rearrangement. It has been suggested that both the extended β-strand and coil segments enhance calcium binding and possibly determine the function of these domains (Collino et al., 2006; Delak et al., 2009a).

One of the most significant features of IDPs is their ability to interact with many different partners (Uversky & Dunker, 2010). Proteins involved in biomineralization have to bind many targets to properly fulfill their functions. They are responsible for macromolecular complex assembly, and they are engaged in cell attachment (Fujisawa et al., 1997; Stubbs et al., 1997), cell signaling (Gruenbaum-Cohen et al., 2009) and/or the regulation of gene expression (Narayanan et al., 2003, 2006). Macromolecular complex assemblies are crucial for proper biomineral formation. The extended conformation of IDPs provides a large binding surface for different partners, while the flexibility and plasticity of polypeptide chains permit IDPs to conformationally adapt to different targets (Collino et al., 2006; Delak et al., 2008). Moreover, proteins involved in biomineralization may undergo disorder-to-order transitions as a result of binding with a partner, molecular crowding or other factors (Amos et al., 2010). Additionally, this disorder-to-order transition promotes macromolecular complex assembly.

Interactions with other macromolecules can also modulate the function of a protein. DPP in solution acts as an inhibitor of HA growth, while DPP bound to collagen acts as a nucleator of HA crystals. This phenomenon might be the result of conformational changes in DPP triggered by interactions with collagen. OPN also exhibits a different conformation when it is bound to HA from that of the solution state (Chien et al., 2009; Gorski et al., 1995). The function of n16 and OMM64 (which is also a putative IDP) is also modulated by interactions with β-chitin. OMM64, n16 and β-chitin in isolation can not induce aragonite formation *in vitro*. However, the combination of OMM46/β-chitin and n16/β-chitin molecules result in aragonite formation. Moreover, it has been suggested that the intrinsic disorder of n16 is a key factor allowing this protein to simultaneously interact with β-chitin and mineral face (Keene et al., 2010; Tohse et al., 2009).

Many proteins involved in biomineralization are able to self-assemble (Kaartinen et al., 1999; Lakshminarayanan et al., 2003). Monomer proteins are often IDPs, while the aggregation of a protein leads to the formation of a β-sheet structure. It is very possible that the extended structure of a protein in solution facilitates interactions with other macromolecules. These interactions may trigger disorder-to-order transitions, which enable macromolecular complexes to form (Buchko et al., 2008; Gorski et al., 1995; He et al., 2003b). Random scrambling of n16 led to a lower affinity for β-chitin and the disappearance of disorder-to-order transitions; this resulted in an inability to form aragonite *in vitro* (Keene et al., 2010). Interestingly, aggregation and conformational change is highly associated with ambient conditions; self-assembly of amelogenin depends on protein concentration, ionic strength, pH and temperature (Lyngstadaas et al., 2009), while DMP1 oligomerization is induced by calcium ions (He et al., 2003b).

A fascinating example of IDPs involved in biomineralization is DMP1. This multifunctional protein is able to interact with calcium ions (He et al., 2003a, 2003b), HA crystals (He et al., 2003a), collagen (Qin et al., 2004), DNA (Narayanan et al., 2006), H factors, integrin αvβ3 and CD44 (Jain et al., 2002). Consequently, DMP1 is engaged in biomineralization, cell signaling and regulation of gene expression (He et al., 2003a; Jain et al., 2002; Narayanan et al., 2006). It exhibits different localization (nuclear, extracellular) according to post-translational modifications and its conformational state (Narayanan et al., 2003). This multifunctionality and interactions with numerous target are likely facilitated by the intrinsically disordered character of the protein.

It is known that proteins involved in biomineralization are extensively post-translationally modified, frequently being phosphorylated, glycosylated and proteolytically cleaved (George & Veis, 2008; He et al., 2005b; Hecker et al., 2003; Qin et al., 2004). A susceptibility to post-translational modifications seems to be associated with the extended and flexible structure that is characteristic of IDPs. The lack of a well-packed hydrophobic core and rapid fluctuations in the peptide chain lead to a greater amount of potential phosphorylation sites, more so than in the case of globular proteins. (Bertrand et al., 1996; Iakoucheva et al., 2004).

Using the IDP classifications (2.2), one can classify the proteins involved in biomineralization into several groups. Those most important for biomineralization are molecular assemblers. This is supported by their ability to interact simultaneously with crystals, scaffold molecules (collagen, β-chitin) and their self-assembling properties. Furthermore, widespread counter ion-binding indicates that they may act as scavengers. We cannot exclude the function of these proteins as effectors, because they are also engaged in processes such as gene expression (Narayanan et al., 2003, 2006). Finally, numerous post-translational modifications suggest the presence of display sites.

It is apparent that intrinsic disorder is characteristic of proteins involved in biomineralization. The structure of many proteins has yet to be established, but there is a considerable likelihood that they are IDPs (Fig. 1B, Table 2).

6. Acknowledgment

This paper was supported in part by the National Science Centre grant 1200/B/H03/2011/40 (to P.D.) and by the Wrocław University of Technology.

7. References

Allemand, D.; Mayer-Gostan, N.; De Pontual, H.; Boeuf, G. & Payan, P. (2007; 2008). Fish Otolith Calcification in Relation to Endolymph Chemistry, Wiley-VCH Verlag GmbH, 9783527619443

Amos, F.F. & Evans, J.S. (2009). AP7, a partially disordered pseudo C-RING protein, is capable of forming stabilized aragonite in vitro. *Biochemistry*, Vol. 48, No. 6, pp. 1332-1339, 1520-4995; 0006-2960

Amos, F.F.; Ndao, M. & Evans, J.S. (2009). Evidence of mineralization activity and supramolecular assembly by the N-terminal sequence of ACCBP, a biomineralization protein that is homologous to the acetylcholine binding protein family. *Biomacromolecules*, Vol. 10, No. 12, pp. 3298-3305, 1526-4602; 1525-7797

Amos, F.F.; Ponce, C.B. & Evans, J.S. (2011). Formation of framework nacre polypeptide supramolecular assemblies that nucleate polymorphs. *Biomacromolecules*, Vol. 12, No. 5, pp. 1883-1890, 1526-4602; 1525-7797

Amos, F.F.; Destine, E.; Ponce, C.B. & Evans, J.S. (2010). The N- and C-Terminal Regions of the Pearl-Associated EF Hand Protein, PFMG1, Promote the Formation of the Aragonite Polymorph in Vitro. *Crystal Growth & Design*, Vol. 10, No. 10, pp. 4211-4216

Arias, J.L.; Mann, K.; Nys, Y.; Ruiz, J.M.G. & Fernández, M.S. (2007; 2008). Eggshell Growth and Matrix Macromolecules, Wiley-VCH Verlag GmbH, 9783527619443

Babu, M.M.; van der Lee, R.; de Groot, N.S. & Gsponer, J. (2011). Intrinsically disordered proteins: regulation and disease. *Current Opinion in Structural Biology*, Vol. 21, No. 3, pp. 432-440, 1879-033X; 0959-440X

Bai, Y.; Chung, J.; Dyson, H.J. & Wright, P.E. (2001). Structural and dynamic characterization of an unfolded state of poplar apo-plastocyanin formed under nondenaturing conditions. *Protein Science: A Publication of the Protein Society*, Vol. 10, No. 5, pp. 1056-1066, 0961-8368; 0961-8368

Bajoghli, B.; Ramialison, M.; Aghaallaei, N.; Czerny, T. & Wittbrodt, J. (2009). Identification of starmaker-like in medaka as a putative target gene of Pax2 in the otic vesicle. *Developmental Dynamics: An Official Publication of the American Association of Anatomists*, Vol. 238, No. 11, pp. 2860-2866, 1097-0177; 1058-8388

Bardet, C.; Vincent, C.; Lajarille, M.C.; Jaffredo, T. & Sire, J.Y. (2010). OC-116, the chicken ortholog of mammalian MEPE found in eggshell, is also expressed in bone cells. *Journal of Experimental Zoology.Part B, Molecular and Developmental Evolution*, Vol. 314, No. 8, pp. 653-662, 1552-5015; 1552-5007

Bartlett, J.D.; Ganss, B.; Goldberg, M.; Moradian-Oldak, J.; Paine, M.L.; Snead, M.L.; Wen, X.; White, S.N. & Zhou, Y.L. (2006). 3. Protein-protein interactions of the developing enamel matrix. *Current Topics in Developmental Biology*, Vol. 74, pp. 57-115, 0070-2153; 0070-2153

Benson, S.; Sucov, H.; Stephens, L.; Davidson, E. & Wilt, F. (1987). A lineage-specific gene encoding a major matrix protein of the sea urchin embryo spicule. I. Authentication of the cloned gene and its developmental expression. *Developmental Biology*, Vol. 120, No. 2, pp. 499-506, 0012-1606; 0012-1606

Bernard, J.P.; Adrich, Z.; Montalto, G.; De Caro, A.; De Reggi, M.; Sarles, H. & Dagorn, J.C. (1992). Inhibition of nucleation and crystal growth of calcium carbonate by human lithostathine. *Gastroenterology*, Vol. 103, No. 4, pp. 1277-1284, 0016-5085; 0016-5085

Bertrand, J.A.; Pignol, D.; Bernard, J.P.; Verdier, J.M.; Dagorn, J.C. & Fontecilla-Camps, J.C. (1996). Crystal structure of human lithostathine, the pancreatic inhibitor of stone formation. *The EMBO Journal*, Vol. 15, No. 11, pp. 2678-2684, 0261-4189; 0261-4189

Bianco, P.; Fisher, L.W.; Young, M.F.; Termine, J.D. & Robey, P.G. (1991). Expression of bone sialoprotein (BSP) in developing human tissues. *Calcified Tissue International*, Vol. 49, No. 6, pp. 421-426, 0171-967X; 0171-967X

Buchko, G.W.; Tarasevich, B.J.; Bekhazi, J.; Snead, M.L. & Shaw, W.J. (2008). A solution NMR investigation into the early events of amelogenin nanosphere self-assembly initiated with sodium chloride or calcium chloride. *Biochemistry*, Vol. 47, No. 50, pp. 13215-13222, 1520-4995; 0006-2960

Buchko, G.W.; Tarasevich, B.J.; Roberts, J.; Snead, M.L. & Shaw, W.J. (2010). A solution NMR investigation into the murine amelogenin splice-variant LRAP (Leucine-Rich Amelogenin Protein). *Biochimica Et Biophysica Acta*, Vol. 1804, No. 9, pp. 1768-1774, 0006-3002; 0006-3002

Chen, Y.; Bal, B.S. & Gorski, J.P. (1992). Calcium and collagen binding properties of osteopontin, bone sialoprotein, and bone acidic glycoprotein-75 from bone. *The Journal of Biological Chemistry*, Vol. 267, No. 34, pp. 24871-24878, 0021-9258; 0021-9258

Chien, Y.C.; Hincke, M.T.; Vali, H. & McKee, M.D. (2008). Ultrastructural matrix-mineral relationships in avian eggshell, and effects of osteopontin on calcite growth in vitro. *Journal of Structural Biology*, Vol. 163, No. 1, pp. 84-99, 1095-8657; 1047-8477

Chien, Y.C.; Masica, D.L.; Gray, J.J.; Nguyen, S.; Vali, H. & McKee, M.D. (2009). Modulation of calcium oxalate dihydrate growth by selective crystal-face binding of phosphorylated osteopontin and polyaspartate peptide showing occlusion by sectoral (compositional) zoning. *The Journal of Biological Chemistry*, Vol. 284, No. 35, pp. 23491-23501, 0021-9258; 0021-9258

Collino, S. & Evans, J.S. (2008). Molecular specifications of a mineral modulation sequence derived from the aragonite-promoting protein n16. *Biomacromolecules*, Vol. 9, No. 7, pp. 1909-1918, 1526-4602; 1525-7797

Collino, S.; Kim, I.W. & Evans, J.S. (2006). Identification of an Acidic C-Terminal Mineral Modification Sequence from the Mollusk Shell Protein Asprich. *Crystal Growth & Design*, Vol. 6, No. 4, pp. 839-842

Cross, K.J.; Huq, N.L. & Reynolds, E.C. (2005). Protein dynamics of bovine dentin phosphophoryn. *The Journal of Peptide Research: Official Journal of the American Peptide Society*, Vol. 66, No. 2, pp. 59-67, 1397-002X; 1397-002X

De Caro, A.; Multigner, L.; Dagorn, J.C. & Sarles, H. (1988). The human pancreatic stone protein. *Biochimie*, Vol. 70, No. 9, pp. 1209-1214, 0300-9084; 0300-9084

Delak, K.; Collino, S. & Evans, J.S. (2009a). Polyelectrolyte domains and intrinsic disorder within the prismatic Asprich protein family. *Biochemistry*, Vol. 48, No. 16, pp. 3669-3677, 1520-4995; 0006-2960

Delak, K.; Harcup, C.; Lakshminarayanan, R.; Sun, Z.; Fan, Y.; Moradian-Oldak, J. & Evans, J.S. (2009b). The tooth enamel protein, porcine amelogenin, is an intrinsically disordered protein with an extended molecular configuration in the monomeric form. *Biochemistry*, Vol. 48, No. 10, pp. 2272-2281, 1520-4995; 0006-2960

Delak, K.; Giocondi, J.; Orme, C. & Evans, J.S. (2008). Modulation of Crystal Growth by the Terminal Sequences of the Prismatic-Associated Asprich Protein. *Crystal Growth & Design*, Vol. 8, No. 12, pp. 4481-4486

Delgado, S.; Girondot, M. & Sire, J.Y. (2005). Molecular evolution of amelogenin in mammals. *Journal of Molecular Evolution*, Vol. 60, No. 1, pp. 12-30, 0022-2844; 0022-2844

Dosztanyi, Z.; Csizmok, V.; Tompa, P. & Simon, I. (2005). IUPred: web server for the prediction of intrinsically unstructured regions of proteins based on estimated energy content. *Bioinformatics (Oxford, England)*, Vol. 21, No. 16, pp. 3433-3434, 1367-4803; 1367-4803

Dunker, A.K.; Cortese, M.S.; Romero, P.; Iakoucheva, L.M. & Uversky, V.N. (2005). Flexible nets. The roles of intrinsic disorder in protein interaction networks. *The FEBS Journal*, Vol. 272, No. 20, pp. 5129-5148, 1742-464X; 1742-464X

Dunker, A.K.; Lawson, J.D.; Brown, C.J.; Williams, R.M.; Romero, P.; Oh, J.S.; Oldfield, C.J.; Campen, A.M.; Ratliff, C.M.; Hipps, K.W.; Ausio, J.; Nissen, M.S.; Reeves, R.; Kang, C.; Kissinger, C.R.; Bailey, R.W.; Griswold, M.D.; Chiu, W.; Garner, E.C. & Obradovic, Z. (2001). Intrinsically disordered protein. *Journal of Molecular Graphics & Modelling*, Vol. 19, No. 1, pp. 26-59, 1093-3263; 1093-3263

Dunker, A.K. & Obradovic, Z. (2001). The protein trinity-linking function and disorder. *Nature Biotechnology*, Vol. 19, No. 9, pp. 805-806, 1087-0156; 1087-0156

Dunker, A.K.; Obradovic, Z.; Romero, P.; Garner, E.C. & Brown, C.J. (2000). Intrinsic protein disorder in complete genomes. *Genome Informatics.Workshop on Genome Informatics*, Vol. 11, pp. 161-171, 0919-9454; 0919-9454

Dyson, H.J. (2011). Expanding the proteome: disordered and alternatively folded proteins. *Quarterly Reviews of Biophysics*, pp. 1-52, 1469-8994; 0033-5835

Dyson, H.J. & Wright, P.E. (2005). Intrinsically unstructured proteins and their functions. *Nature Reviews. Molecular Cell Biology*, Vol. 6, No. 3, pp. 197-208, 1471-0072; 1471-0072

Endo, H.; Takagi, Y.; Ozaki, N.; Kogure, T. & Watanabe, T. (2004). A crustacean Ca^{2+}-binding protein with a glutamate-rich sequence promotes $CaCO_3$ crystallization. *The Biochemical Journal*, Vol. 384, No. Pt 1, pp. 159-167, 1470-8728; 0264-6021

Evans, J.S. (2008). "Tuning in" to mollusk shell nacre- and prismatic-associated protein terminal sequences. Implications for biomineralization and the construction of high performance inorganic-organic composites. *Chemical Reviews*, Vol. 108, No. 11, pp. 4455-4462, 1520-6890; 0009-2665

Evans, J.S.; Chiu, T. & Chan, S.I. (1994). Phosphophoryn, an "acidic" biomineralization regulatory protein: conformational folding in the presence of Cd(II). *Biopolymers*, Vol. 34, No. 10, pp. 1359-1375, 0006-3525; 0006-3525

Fisher, L.W.; Torchia, D.A.; Fohr, B.; Young, M.F. & Fedarko, N.S. (2001). Flexible structures of SIBLING proteins, bone sialoprotein, and osteopontin. *Biochemical and Biophysical Research Communications*, Vol. 280, No. 2, pp. 460-465, 0006-291X; 0006-291X

Fu, G.; Valiyaveettil, S.; Wopenka, B. & Morse, D.E. (2005). $CaCO_3$ biomineralization: acidic 8-kDa proteins isolated from aragonitic abalone shell nacre can specifically modify calcite crystal morphology. *Biomacromolecules*, Vol. 6, No. 3, pp. 1289-1298, 1525-7797; 1525-7797

Fujisawa, R. & Kuboki, Y. (1998). Conformation of dentin phosphophoryn adsorbed on hydroxyapatite crystals. *European Journal of Oral Sciences*, Vol. 106 Suppl 1, pp. 249-253, 0909-8836; 0909-8836

Fujisawa, R.; Mizuno, M.; Nodasaka, Y. & Kuboki, Y. (1997). Attachment of osteoblastic cells to hydroxyapatite crystals by a synthetic peptide (Glu7-Pro-Arg-Gly-Asp-Thr) containing two functional sequences of bone sialoprotein. *Matrix Biology: Journal of the International Society for Matrix Biology*, Vol. 16, No. 1, pp. 21-28, 0945-053X; 0945-053X

Ganss, B.; Kim, R.H. & Sodek, J. (1999). Bone sialoprotein. *Critical Reviews in Oral Biology and Medicine : An Official Publication of the American Association of Oral Biologists*, Vol. 10, No. 1, pp. 79-98, 1045-4411; 1045-4411

Geider, S.; Baronnet, A.; Cerini, C.; Nitsche, S.; Astier, J.P.; Michel, R.; Boistelle, R.; Berland, Y.; Dagorn, J.C. & Verdier, J.M. (1996). Pancreatic lithostathine as a calcite habit modifier. *The Journal of Biological Chemistry*, Vol. 271, No. 42, pp. 26302-26306, 0021-9258; 0021-9258

George, A. & Hao, J. (2005). Role of phosphophoryn in dentin mineralization. *Cells, Tissues, Organs*, Vol. 181, No. 3-4, pp. 232-240, 1422-6405; 1422-6405

George, A. & Veis, A. (2008). Phosphorylated proteins and control over apatite nucleation, crystal growth, and inhibition. *Chemical Reviews*, Vol. 108, No. 11, pp. 4670-4693, 1520-6890; 0009-2665

George, N.C.; Killian, C.E. & Wilt, F.H. (1991). Characterization and expression of a gene encoding a 30.6-kDa Strongylocentrotus purpuratus spicule matrix protein. *Developmental Biology*, Vol. 147, No. 2, pp. 334-342, 0012-1606; 0012-1606

Gerbaud, V.; Pignol, D.; Loret, E.; Bertrand, J.A.; Berland, Y.; Fontecilla-Camps, J.C.; Canselier, J.P.; Gabas, N. & Verdier, J.M. (2000). Mechanism of calcite crystal growth inhibition by the N-terminal undecapeptide of lithostathine. *The Journal of Biological Chemistry*, Vol. 275, No. 2, pp. 1057-1064, 0021-9258; 0021-9258

Gericke, A.; Qin, C.; Spevak, L.; Fujimoto, Y.; Butler, W.T.; Sorensen, E.S. & Boskey, A.L. (2005). Importance of phosphorylation for osteopontin regulation of biomineralization. *Calcified Tissue International*, Vol. 77, No. 1, pp. 45-54, 0171-967X; 0171-967X

Gericke, A.; Qin, C.; Sun, Y.; Redfern, R.; Redfern, D.; Fujimoto, Y.; Taleb, H.; Butler, W.T. & Boskey, A.L. (2010). Different forms of DMP1 play distinct roles in mineralization. *Journal of Dental Research*, Vol. 89, No. 4, pp. 355-359, 1544-0591; 0022-0345

Glazer, L.; Shechter, A.; Tom, M.; Yudkovski, Y.; Weil, S.; Aflalo, E.D.; Pamuru, R.R.; Khalaila, I.; Bentov, S.; Berman, A. & Sagi, A. (2010). A protein involved in the assembly of an extracellular calcium storage matrix. *The Journal of Biological Chemistry*, Vol. 285, No. 17, pp. 12831-12839, 1083-351X; 0021-9258

Glimcher, M.J. (1989). Mechanism of calcification: role of collagen fibrils and collagen-phosphoprotein complexes in vitro and in vivo. *The Anatomical Record*, Vol. 224, No. 2, pp. 139-153, 0003-276X; 0003-276X

Goldberg, H.A. & Hunter, G.K. (1995). The inhibitory activity of osteopontin on hydroxyapatite formation in vitro. *Annals of the New York Academy of Sciences*, Vol. 760, pp. 305-308, 0077-8923; 0077-8923

Gorski, J.P.; Kremer, E.; Ruiz-Perez, J.; Wise, G.E. & Artigues, A. (1995). Conformational analyses on soluble and surface bound osteopontin. *Annals of the New York Academy of Sciences*, Vol. 760, pp. 12-23, 0077-8923; 0077-8923

Gotliv, B.A.; Kessler, N.; Sumerel, J.L.; Morse, D.E.; Tuross, N.; Addadi, L. & Weiner, S. (2005). Asprich: A novel aspartic acid-rich protein family from the prismatic shell matrix of the bivalve Atrina rigida. *Chembiochem : A European Journal of Chemical Biology*, Vol. 6, No. 2, pp. 304-314, 1439-4227; 1439-4227

Grohe, B.; O'Young, J.; Ionescu, D.A.; Lajoie, G.; Rogers, K.A.; Karttunen, M.; Goldberg, H.A. & Hunter, G.K. (2007). Control of calcium oxalate crystal growth by face-specific adsorption of an osteopontin phosphopeptide. *Journal of the American Chemical Society*, Vol. 129, No. 48, pp. 14946-14951, 1520-5126; 0002-7863

Gruenbaum-Cohen, Y.; Tucker, A.S.; Haze, A.; Shilo, D.; Taylor, A.L.; Shay, B.; Sharpe, P.T.; Mitsiadis, T.A.; Ornoy, A.; Blumenfeld, A. & Deutsch, D. (2009). Amelogenin in cranio-facial development: the tooth as a model to study the role of amelogenin during embryogenesis. *Journal of Experimental Zoology.Part B, Molecular and Developmental Evolution*, Vol. 312B, No. 5, pp. 445-457, 1552-5015; 1552-5007

Harkey, M.A.; Klueg, K.; Sheppard, P. & Raff, R.A. (1995). Structure, expression, and extracellular targeting of PM27, a skeletal protein associated specifically with growth of the sea urchin larval spicule. *Developmental Biology*, Vol. 168, No. 2, pp. 549-566, 0012-1606; 0012-1606

Hay, D.I.; Smith, D.J.; Schluckebier, S.K. & Moreno, E.C. (1984). Relationship between concentration of human salivary statherin and inhibition of calcium phosphate precipitation in stimulated human parotid saliva. *Journal of Dental Research*, Vol. 63, No. 6, pp. 857-863, 0022-0345; 0022-0345

He, G.; Dahl, T.; Veis, A. & George, A. (2003a). Dentin matrix protein 1 initiates hydroxyapatite formation in vitro. *Connective Tissue Research*, Vol. 44 Suppl 1, pp. 240-245, 0300-8207; 0300-8207

He, G.; Dahl, T.; Veis, A. & George, A. (2003b). Nucleation of apatite crystals in vitro by self-assembled dentin matrix protein 1. *Nature Materials*, Vol. 2, No. 8, pp. 552-558, 1476-1122; 1476-1122

He, G.; Gajjeraman, S.; Schultz, D.; Cookson, D.; Qin, C.; Butler, W.T.; Hao, J. & George, A. (2005a). Spatially and temporally controlled biomineralization is facilitated by interaction between self-assembled dentin matrix protein 1 and calcium phosphate nuclei in solution. *Biochemistry*, Vol. 44, No. 49, pp. 16140-16148, 0006-2960; 0006-2960

He, G.; Ramachandran, A.; Dahl, T.; George, S.; Schultz, D.; Cookson, D.; Veis, A. & George, A. (2005b). Phosphorylation of phosphophoryn is crucial for its function as a mediator of biomineralization. *The Journal of Biological Chemistry*, Vol. 280, No. 39, pp. 33109-33114, 0021-9258; 0021-9258

Hecker, A.; Quennedey, B.; Testeniere, O.; Quennedey, A.; Graf, F. & Luquet, G. (2004). Orchestin, a calcium-binding phosphoprotein, is a matrix component of two successive transitory calcified biomineralizations cyclically elaborated by a terrestrial crustacean. *Journal of Structural Biology*, Vol. 146, No. 3, pp. 310-324, 1047-8477; 1047-8477

Hecker, A.; Testeniere, O.; Marin, F. & Luquet, G. (2003). Phosphorylation of serine residues is fundamental for the calcium-binding ability of Orchestin, a soluble matrix protein from crustacean calcium storage structures. *FEBS Letters*, Vol. 535, No. 1-3, pp. 49-54, 0014-5793; 0014-5793

Hincke, M.T.; Gautron, J.; Tsang, C.P.; McKee, M.D. & Nys, Y. (1999). Molecular cloning and ultrastructural localization of the core protein of an eggshell matrix proteoglycan, ovocleidin-116. *The Journal of Biological Chemistry*, Vol. 274, No. 46, pp. 32915-32923, 0021-9258; 0021-9258

Hincke, M.T.; Tsang, C.P.; Courtney, M.; Hill, V. & Narbaitz, R. (1995). Purification and immunochemistry of a soluble matrix protein of the chicken eggshell (ovocleidin 17). *Calcified Tissue International*, Vol. 56, No. 6, pp. 578-583, 0171-967X; 0171-967X

Huq, N.L.; Cross, K.J.; Talbo, G.H.; Riley, P.F.; Loganathan, A.; Crossley, M.A.; Perich, J.W. & Reynolds, E.C. (2000). N-terminal sequence analysis of bovine dentin phosphophoryn after conversion of phosphoseryl to S-propylcysteinyl residues. *Journal of Dental Research*, Vol. 79, No. 11, pp. 1914-1919, 0022-0345; 0022-0345

Iakoucheva, L.M.; Radivojac, P.; Brown, C.J.; O'Connor, T.R.; Sikes, J.G.; Obradovic, Z. & Dunker, A.K. (2004). The importance of intrinsic disorder for protein phosphorylation. *Nucleic Acids Research*, Vol. 32, No. 3, pp. 1037-1049, 1362-4962; 0305-1048

Illies, M.R.; Peeler, M.T.; Dechtiaruk, A.M. & Ettensohn, C.A. (2002). Identification and developmental expression of new biomineralization proteins in the sea urchin Strongylocentrotus purpuratus. *Development Genes and Evolution*, Vol. 212, No. 9, pp. 419-431, 0949-944X; 0949-944X

Inoue, H.; Ohira, T.; Ozaki, N. & Nagasawa, H. (2004). A novel calcium-binding peptide from the cuticle of the crayfish, Procambarus clarkii. *Biochemical and Biophysical Research Communications*, Vol. 318, No. 3, pp. 649-654, 0006-291X; 0006-291X

Inoue, H.; Ohira, T.; Ozaki, N. & Nagasawa, H. (2003). Cloning and expression of a cDNA encoding a matrix peptide associated with calcification in the exoskeleton of the crayfish. *Comparative Biochemistry and Physiology.Part B, Biochemistry & Molecular Biology*, Vol. 136, No. 4, pp. 755-765, 1096-4959; 1096-4959

Inoue, H.; Ozaki, N. & Nagasawa, H. (2001). Purification and structural determination of a phosphorylated peptide with anti-calcification and chitin-binding activities in the exoskeleton of the crayfish, Procambarus clarkii. *Bioscience, Biotechnology, and Biochemistry*, Vol. 65, No. 8, pp. 1840-1848, 0916-8451; 0916-8451

Inoue, H.; Yuasa-Hashimoto, N.; Suzuki, M. & Nagasawa, H. (2008). Structural determination and functional analysis of a soluble matrix protein associated with calcification of the exoskeleton of the crayfish, Procambarus clarkii. *Bioscience, Biotechnology, and Biochemistry*, Vol. 72, No. 10, pp. 2697-2707, 1347-6947; 0916-8451

Inoue, H.; Ohira, T. & Nagasawa, H. (2007). Significance of the N- and C-terminal regions of CAP-1, a cuticle calcification-associated peptide from the exoskeleton of the crayfish, for calcification. *Peptides*, Vol. 28, No. 3, pp. 566-573, 0196-9781

Ishii, K.; Yanagisawa, T. & Nagasawa, H. (1996). Characterization of a matrix protein in the gastroliths of the crayfish Procambarus clarkii. *Bioscience, Biotechnology, and Biochemistry*, Vol. 60, No. 9, pp. 1479-1482, 0916-8451; 0916-8451

Jain, A.; Karadag, A.; Fohr, B.; Fisher, L.W. & Fedarko, N.S. (2002). Three SIBLINGs (small integrin-binding ligand, N-linked glycoproteins) enhance factor H's cofactor activity enabling MCP-like cellular evasion of complement-mediated attack. *The Journal of Biological Chemistry*, Vol. 277, No. 16, pp. 13700-13708, 0021-9258; 0021-9258

Johnsson, M.; Richardson, C.F.; Bergey, E.J.; Levine, M.J. & Nancollas, G.H. (1991). The effects of human salivary cystatins and statherin on hydroxyapatite crystallization. *Archives of Oral Biology*, Vol. 36, No. 9, pp. 631-636, 0003-9969; 0003-9969

Jono, S.; Peinado, C. & Giachelli, C.M. (2000). Phosphorylation of osteopontin is required for inhibition of vascular smooth muscle cell calcification. *The Journal of Biological Chemistry*, Vol. 275, No. 26, pp. 20197-20203, 0021-9258; 0021-9258

Kaartinen, M.T.; Pirhonen, A.; Linnala-Kankkunen, A. & Maenpaa, P.H. (1999). Cross-linking of osteopontin by tissue transglutaminase increases its collagen binding properties. *The Journal of Biological Chemistry*, Vol. 274, No. 3, pp. 1729-1735, 0021-9258; 0021-9258

Kaplon, T.M.; Michnik, A.; Drzazga, Z.; Richter, K.; Kochman, M. & Ozyhar, A. (2009). The rod-shaped conformation of Starmaker. *Biochimica et Biophysica Acta*, Vol. 1794, No. 11, pp. 1616-1624, 0006-3002; 0006-3002

Kaplon, T.M.; Rymarczyk, G.; Nocula-Lugowska, M.; Jakob, M.; Kochman, M.; Lisowski, M.; Szewczuk, Z. & Ozyhar, A. (2008). Starmaker exhibits properties of an intrinsically disordered protein. *Biomacromolecules*, Vol. 9, No. 8, pp. 2118-2125, 1526-4602; 1525-7797

Katoh-Fukui, Y.; Noce, T.; Ueda, T.; Fujiwara, Y.; Hashimoto, N.; Higashinakagawa, T.; Killian, C.E.; Livingston, B.T.; Wilt, F.H. & Benson, S.C. (1991). The corrected structure of the SM50 spicule matrix protein of Strongylocentrotus purpuratus. *Developmental Biology*, Vol. 145, No. 1, pp. 201-202, 0012-1606; 0012-1606

Kazanecki, C.C.; Uzwiak, D.J. & Denhardt, D.T. (2007). Control of osteopontin signaling and function by post-translational phosphorylation and protein folding. *Journal of Cellular Biochemistry*, Vol. 102, No. 4, pp. 912-924, 0730-2312; 0730-2312

Keene, E.C.; Evans, J.S. & Estroff, L.A. (2010). Matrix Interactions in Biomineralization: Aragonite Nucleation by an Intrinsically Disordered Nacre Polypeptide, n16N, Associated with a β-Chitin Substrate. *Crystal Growth & Design*, Vol. 10, No. 3, pp. 1383-1389

Killian, C.E.; Croker, L. & Wilt, F.H. (2010). SpSM30 gene family expression patterns in embryonic and adult biomineralized tissues of the sea urchin, Strongylocentrotus purpuratus. *Gene Expression Patterns: GEP*, Vol. 10, No. 2-3, pp. 135-139, 1872-7298; 1567-133X

Killian, C.E. & Wilt, F.H. (2008). Molecular aspects of biomineralization of the echinoderm endoskeleton. *Chemical Reviews*, Vol. 108, No. 11, pp. 4463-4474, 1520-6890; 0009-2665

Killian, C.E. & Wilt, F.H. (1996). Characterization of the proteins comprising the integral matrix of Strongylocentrotus purpuratus embryonic spicules. *The Journal of Biological Chemistry*, Vol. 271, No. 15, pp. 9150-9159, 0021-9258; 0021-9258

Kim, I.W.; Morse, D.E. & Evans, J.S. (2004). Molecular characterization of the 30-AA N-terminal mineral interaction domain of the biomineralization protein AP7.

Langmuir: The ACS Journal of Surfaces and Colloids, Vol. 20, No. 26, pp. 11664-11673, 0743-7463; 0743-7463

Kim, I.W.; Collino, S.; Morse, D.E. & Evans, J.S. (2006a). A Crystal Modulating Protein from Molluscan Nacre That Limits the Growth of Calcite in Vitro. *Crystal Growth & Design*, Vol. 6, No. 5, pp. 1078-1082

Kim, I.W.; Darragh, M.R.; Orme, C. & Evans, J.S. (2006b). Molecular Tuning of Crystal Growth by Nacre-Associated Polypeptides. *Crystal Growth & Design*, Vol. 6, No. 1, pp. 5-10

Kim, I.W.; DiMasi, E. & Evans, J.S. (2004). Identification of Mineral Modulation Sequences within the Nacre-Associated Oyster Shell Protein, n16. *Crystal Growth & Design*, Vol. 4, No. 6, pp. 1113-1118

Kim, I.W.; Giocondi, J.L.; Orme, C.; Collino, S. & Spencer Evans, J. (2008). Morphological and Kinetic Transformation of Calcite Crystal Growth by Prismatic-Associated Asprich Sequences. *Crystal Growth & Design*, Vol. 8, No. 4, pp. 1154-1160

Lakshminarayanan, R.; Joseph, J.S.; Kini, R.M. & Valiyaveettil, S. (2005). Structure-function relationship of avian eggshell matrix proteins: a comparative study of two major eggshell matrix proteins, ansocalcin and OC-17. *Biomacromolecules*, Vol. 6, No. 2, pp. 741-751, 1525-7797; 1525-7797

Lakshminarayanan, R.; Valiyaveettil, S.; Rao, V.S. & Kini, R.M. (2003). Purification, characterization, and in vitro mineralization studies of a novel goose eggshell matrix protein, ansocalcin. *The Journal of Biological Chemistry*, Vol. 278, No. 5, pp. 2928-2936, 0021-9258; 0021-9258

Lee, S.L.; Veis, A. & Glonek, T. (1977). Dentin phosphoprotein: an extracellular calcium-binding protein. *Biochemistry*, Vol. 16, No. 13, pp. 2971-2979

Lee, Y.H.; Britten, R.J. & Davidson, E.H. (1999). SM37, a skeletogenic gene of the sea urchin embryo linked to the SM50 gene. *Development, Growth & Differentiation*, Vol. 41, No. 3, pp. 303-312, 0012-1592; 0012-1592

Li, M.; Liu, J.; Ran, X.; Fang, M.; Shi, J.; Qin, H.; Goh, J.M. & Song, J. (2006). Resurrecting abandoned proteins with pure water: CD and NMR studies of protein fragments solubilized in salt-free water. *Biophysical Journal*, Vol. 91, No. 11, pp. 4201-4209, 0006-3495; 0006-3495

Linding, R.; Schymkowitz, J.; Rousseau, F.; Diella, F. & Serrano, L. (2004). A comparative study of the relationship between protein structure and beta-aggregation in globular and intrinsically disordered proteins. *Journal of Molecular Biology*, Vol. 342, No. 1, pp. 345-353, 0022-2836; 0022-2836

Liu, H.L.; Liu, S.F.; Ge, Y.J.; Liu, J.; Wang, X.Y.; Xie, L.P.; Zhang, R.Q. & Wang, Z. (2007). Identification and characterization of a biomineralization related gene PFMG1 highly expressed in the mantle of Pinctada fucata. *Biochemistry*, Vol. 46, No. 3, pp. 844-851, 0006-2960; 0006-2960

Livingston, B.T.; Killian, C.E.; Wilt, F.; Cameron, A.; Landrum, M.J.; Ermolaeva, O.; Sapojnikov, V.; Maglott, D.R.; Buchanan, A.M. & Ettensohn, C.A. (2006). A genome-wide analysis of biomineralization-related proteins in the sea urchin Strongylocentrotus purpuratus. *Developmental Biology*, Vol. 300, No. 1, pp. 335-348, 0012-1606; 0012-1606

Long, J.R.; Shaw, W.J.; Stayton, P.S. & Drobny, G.P. (2001). Structure and dynamics of hydrated statherin on hydroxyapatite as determined by solid-state NMR. *Biochemistry*, Vol. 40, No. 51, pp. 15451-15455, 0006-2960; 0006-2960

Luquet, G.; Testeniere, O. & Graf, F. (1996). Characterization and N-terminal sequencing of a calcium binding protein from the calcareous concretion organic matrix of the terrestrial crustacean Orchestia cavimana. *Biochimica et Biophysica Acta*, Vol. 1293, No. 2, pp. 272-276, 0006-3002; 0006-3002

Lyngstadaas, S.P.; Wohlfahrt, J.C.; Brookes, S.J.; Paine, M.L.; Snead, M.L. & Reseland, J.E. (2009). Enamel matrix proteins; old molecules for new applications. *Orthodontics & Craniofacial Research*, Vol. 12, No. 3, pp. 243-253, 1601-6343; 1601-6335

Ma, Z.; Huang, J.; Sun, J.; Wang, G.; Li, C.; Xie, L. & Zhang, R. (2007). A novel extrapallial fluid protein controls the morphology of nacre lamellae in the pearl oyster, Pinctada fucata. *The Journal of Biological Chemistry*, Vol. 282, No. 32, pp. 23253-23263, 0021-9258; 0021-9258

Mann, K. (2004). Identification of the major proteins of the organic matrix of emu (Dromaius novaehollandiae) and rhea (Rhea americana) eggshell calcified layer. *British Poultry Science*, Vol. 45, No. 4, pp. 483-490, 0007-1668; 0007-1668

Mann, K.; Hincke, M.T. & Nys, Y. (2002). Isolation of ovocleidin-116 from chicken eggshells, correction of its amino acid sequence and identification of disulfide bonds and glycosylated Asn. *Matrix Biology: Journal of the International Society for Matrix Biology*, Vol. 21, No. 5, pp. 383-387, 0945-053X; 0945-053X

Mann, K.; Poustka, A.J. & Mann, M. (2010). Phosphoproteomes of Strongylocentrotus purpuratus shell and tooth matrix: identification of a major acidic sea urchin tooth phosphoprotein, phosphodontin. *Proteome Science*, Vol. 8, No. 1, pp. 6, 1477-5956; 1477-5956

Mann, K. & Siedler, F. (2006). Amino acid sequences and phosphorylation sites of emu and rhea eggshell C-type lectin-like proteins. *Comparative Biochemistry and Physiology.Part B, Biochemistry & Molecular Biology*, Vol. 143, No. 2, pp. 160-170, 1096-4959; 1096-4959

Mann, K. & Siedler, F. (2004). Ostrich (Struthio camelus) eggshell matrix contains two different C-type lectin-like proteins. Isolation, amino acid sequence, and posttranslational modifications. *Biochimica et Biophysica Acta*, Vol. 1696, No. 1, pp. 41-50, 0006-3002; 0006-3002

Marie, B.; Luquet, G.; Bedouet, L.; Milet, C.; Guichard, N.; Medakovic, D. & Marin, F. (2008). Nacre calcification in the freshwater mussel Unio pictorum: carbonic anhydrase activity and purification of a 95 kDa calcium-binding glycoprotein. *Chembiochem: A European Journal of Chemical Biology*, Vol. 9, No. 15, pp. 2515-2523, 1439-7633; 1439-4227

Marie, B.; Zanella-Cleon, I.; Corneillat, M.; Becchi, M.; Alcaraz, G.; Plasseraud, L.; Luquet, G. & Marin, F. (2011). Nautilin-63, a novel acidic glycoprotein from the shell nacre of Nautilus macromphalus. *The FEBS Journal*, 1742-4658; 1742-464X

Marin, F.; Amons, R.; Guichard, N.; Stigter, M.; Hecker, A.; Luquet, G.; Layrolle, P.; Alcaraz, G.; Riondet, C. & Westbroek, P. (2005). Caspartin and calprismin, two proteins of the shell calcitic prisms of the Mediterranean fan mussel Pinna nobilis. *The Journal of Biological Chemistry*, Vol. 280, No. 40, pp. 33895-33908, 0021-9258; 0021-9258

Mendoza, C.; Figueirido, F. & Tasayco, M.L. (2003). DSC studies of a family of natively disordered fragments from Escherichia coli thioredoxin: surface burial in intrinsic coils. *Biochemistry*, Vol. 42, No. 11, pp. 3349-3358, 0006-2960; 0006-2960

Meyran, J.; Graf, F. & Nicaise, G. (1984). Calcium pathway through a mineralizing epithelium in the crustacean Orchestia in pre-molt: Ultrastructural cytochemistry and X-ray microanalysis. *Tissue and Cell*, Vol. 16, No. 2, pp. 269-286, 0040-8166

Michenfelder, M.; Fu, G.; Lawrence, C.; Weaver, J.C.; Wustman, B.A.; Taranto, L.; Evans, J.S. & Morse, D.E. (2003). Characterization of two molluscan crystal-modulating biomineralization proteins and identification of putative mineral binding domains. *Biopolymers*, Vol. 70, No. 4, pp. 522-533, 0006-3525; 0006-3525

Millett, I.S.; Doniach, S. & Plaxco, K.W. (2002). Toward a taxonomy of the denatured state: small angle scattering studies of unfolded proteins. *Advances in Protein Chemistry*, Vol. 62, pp. 241-262, 0065-3233; 0065-3233

Moore, E.W. & Verine, H.J. (1987). Pancreatic calcification and stone formation: a thermodynamic model of calcium in pancreatic juice. *The American Journal of Physiology*, Vol. 252, No. 5 Pt 1, pp. G707-18, 0002-9513; 0002-9513

Naganagowda, G.A.; Gururaja, T.L. & Levine, M.J. (1998). Delineation of conformational preferences in human salivary statherin by 1H, 31P NMR and CD studies: sequential assignment and structure-function correlations. *Journal of Biomolecular Structure & Dynamics*, Vol. 16, No. 1, pp. 91-107, 0739-1102; 0739-1102

Narayanan, K.; Gajjeraman, S.; Ramachandran, A.; Hao, J. & George, A. (2006). Dentin matrix protein 1 regulates dentin sialophosphoprotein gene transcription during early odontoblast differentiation. *The Journal of Biological Chemistry*, Vol. 281, No. 28, pp. 19064-19071, 0021-9258; 0021-9258

Narayanan, K.; Ramachandran, A.; Hao, J.; He, G.; Park, K.W.; Cho, M. & George, A. (2003). Dual functional roles of dentin matrix protein 1. Implications in biomineralization and gene transcription by activation of intracellular Ca2+ store. *The Journal of Biological Chemistry*, Vol. 278, No. 19, pp. 17500-17508, 0021-9258; 0021-9258

Ndao, M.; Dutta, K.; Bromley, K.M.; Lakshminarayanan, R.; Sun, Z.; Rewari, G.; Moradian-Oldak, J. & Evans, J.S. (2011). Probing the self-association, intermolecular contacts, and folding propensity of amelogenin. *Protein Science : A Publication of the Protein Society*, Vol. 20, No. 4, pp. 724-734, 1469-896X; 0961-8368

Ndao, M.; Keene, E.; Amos, F.F.; Rewari, G.; Ponce, C.B.; Estroff, L. & Evans, J.S. (2010). Intrinsically disordered mollusk shell prismatic protein that modulates calcium carbonate crystal growth. *Biomacromolecules*, Vol. 11, No. 10, pp. 2539-2544, 1526-4602; 1525-7797

Oldberg, A.; Franzen, A. & Heinegard, D. (1988). The primary structure of a cell-binding bone sialoprotein. *The Journal of Biological Chemistry*, Vol. 263, No. 36, pp. 19430-19432, 0021-9258; 0021-9258

Panheleux, M.; Bain, M.; Fernandez, M.S.; Morales, I.; Gautron, J.; Arias, J.L.; Solomon, S.E.; Hincke, M. & Nys, Y. (1999). Organic matrix composition and ultrastructure of eggshell: a comparative study. *British Poultry Science*, Vol. 40, No. 2, pp. 240-252, 0007-1668; 0007-1668

Patthy, L. (1988). Homology of human pancreatic stone protein with animal lectins. *The Biochemical Journal*, Vol. 253, No. 1, pp. 309-311, 0264-6021; 0264-6021

Pauling, L. & Delbruck, M. (1940). The Nature of the Intermolecular Forces Operative in Biological Processes. *Science (New York, N.Y.)*, Vol. 92, No. 2378, pp. 77-79, 0036-8075; 0036-8075

Prince, C.W.; Dickie, D. & Krumdieck, C.L. (1991). Osteopontin, a substrate for transglutaminase and factor XIII activity. *Biochemical and Biophysical Research Communications*, Vol. 177, No. 3, pp. 1205-1210, 0006-291X; 0006-291X

Qin, C.; Baba, O. & Butler, W.T. (2004). Post-translational modifications of sibling proteins and their roles in osteogenesis and dentinogenesis. *Critical Reviews in Oral Biology and Medicine : An Official Publication of the American Association of Oral Biologists*, Vol. 15, No. 3, pp. 126-136, 1544-1113; 1045-4411

Raj, P.A.; Johnsson, M.; Levine, M.J. & Nancollas, G.H. (1992). Salivary statherin. Dependence on sequence, charge, hydrogen bonding potency, and helical conformation for adsorption to hydroxyapatite and inhibition of mineralization. *The Journal of Biological Chemistry*, Vol. 267, No. 9, pp. 5968-5976, 0021-9258; 0021-9258

Reyes-Grajeda, J.P.; Moreno, A. & Romero, A. (2004). Crystal structure of ovocleidin-17, a major protein of the calcified Gallus gallus eggshell: implications in the calcite mineral growth pattern. *The Journal of Biological Chemistry*, Vol. 279, No. 39, pp. 40876-40881, 0021-9258; 0021-9258

Romero, P.; Obradovic, Z.; Li, X.; Garner, E.C.; Brown, C.J. & Dunker, A.K. (2001). Sequence complexity of disordered protein. *Proteins*, Vol. 42, No. 1, pp. 38-48, 0887-3585; 0887-3585

Ross, M.D. & Pote, K.G. (1984). Some Properties of Otoconia. *Philosophical Transactions of the Royal Society of London.B, Biological Sciences*, Vol. 304, No. 1121, pp. 445-452

Samata, T.; Ikeda, D.; Kajikawa, A.; Sato, H.; Nogawa, C.; Yamada, D.; Yamazaki, R. & Akiyama, T. (2008). A novel phosphorylated glycoprotein in the shell matrix of the oyster Crassostrea nippona. *The FEBS Journal*, Vol. 275, No. 11, pp. 2977-2989, 1742-464X; 1742-464X

Samata, T.; Hayashi, N.; Kono, M.; Hasegawa, K.; Horita, C. & Akera, S. (1999). A new matrix protein family related to the nacreous layer formation of Pinctada fucata. *FEBS Letters*, Vol. 462, No. 1-2, pp. 225-229, 0014-5793

Sarashina, I. & Endo, K. (2001). The complete primary structure of molluscan shell protein 1 (MSP-1), an acidic glycoprotein in the shell matrix of the scallop Patinopecten yessoensis. *Marine Biotechnology (New York, N.Y.)*, Vol. 3, No. 4, pp. 362-369, 1436-2228; 1436-2228

Shapiro, J.L.; Wen, X.; Okamoto, C.T.; Wang, H.J.; Lyngstadaas, S.P.; Goldberg, M.; Snead, M.L. & Paine, M.L. (2007). Cellular uptake of amelogenin, and its localization to CD63, and Lamp1-positive vesicles. *Cellular and Molecular Life Sciences: CMLS*, Vol. 64, No. 2, pp. 244-256, 1420-682X; 1420-682X

Shaw, W.J.; Ferris, K.; Tarasevich, B. & Larson, J.L. (2008). The structure and orientation of the C-terminus of LRAP. *Biophysical Journal*, Vol. 94, No. 8, pp. 3247-3257, 1542-0086; 0006-3495

Shechter, A.; Glazer, L.; Cheled, S.; Mor, E.; Weil, S.; Berman, A.; Bentov, S.; Aflalo, E.D.; Khalaila, I. & Sagi, A. (2008). A gastrolith protein serving a dual role in the formation of an amorphous mineral containing extracellular matrix. *Proceedings of the National Academy of Sciences of the United States of America*, Vol. 105, No. 20, pp. 7129-7134, 1091-6490; 0027-8424

Shen, X.; Belcher, A.M.; Hansma, P.K.; Stucky, G.D. & Morse, D.E. (1997). Molecular cloning and characterization of lustrin A, a matrix protein from shell and pearl nacre of Haliotis rufescens. *The Journal of Biological Chemistry*, Vol. 272, No. 51, pp. 32472-32481, 0021-9258; 0021-9258

Sickmeier, M.; Hamilton, J.A.; LeGall, T.; Vacic, V.; Cortese, M.S.; Tantos, A.; Szabo, B.; Tompa, P.; Chen, J.; Uversky, V.N.; Obradovic, Z. & Dunker, A.K. (2007). DisProt: the Database of Disordered Proteins. *Nucleic Acids Research*, Vol. 35, No. Database issue, pp. D786-93, 1362-4962; 0305-1048

Sollner, C.; Burghammer, M.; Busch-Nentwich, E.; Berger, J.; Schwarz, H.; Riekel, C. & Nicolson, T. (2003). Control of crystal size and lattice formation by starmaker in otolith biomineralization. *Science (New York, N.Y.)*, Vol. 302, No. 5643, pp. 282-286, 1095-9203; 0036-8075

Stayton, P.S.; Drobny, G.P.; Shaw, W.J.; Long, J.R. & Gilbert, M. (2003). Molecular recognition at the protein-hydroxyapatite interface. *Critical Reviews in Oral Biology and Medicine : An Official Publication of the American Association of Oral Biologists*, Vol. 14, No. 5, pp. 370-376, 1544-1113; 1045-4411

Steitz, S.A.; Speer, M.Y.; McKee, M.D.; Liaw, L.; Almeida, M.; Yang, H. & Giachelli, C.M. (2002). Osteopontin inhibits mineral deposition and promotes regression of ectopic calcification. *The American Journal of Pathology*, Vol. 161, No. 6, pp. 2035-2046, 0002-9440; 0002-9440

Stubbs, J.T.,3rd; Mintz, K.P.; Eanes, E.D.; Torchia, D.A. & Fisher, L.W. (1997). Characterization of native and recombinant bone sialoprotein: delineation of the mineral-binding and cell adhesion domains and structural analysis of the RGD domain. *Journal of Bone and Mineral Research : The Official Journal of the American Society for Bone and Mineral Research*, Vol. 12, No. 8, pp. 1210-1222, 0884-0431; 0884-0431

Suzuki, M.; Murayama, E.; Inoue, H.; Ozaki, N.; Tohse, H.; Kogure, T. & Nagasawa, H. (2004). Characterization of Prismalin-14, a novel matrix protein from the prismatic layer of the Japanese pearl oyster (Pinctada fucata). *The Biochemical Journal*, Vol. 382, No. Pt 1, pp. 205-213, 1470-8728; 0264-6021

Suzuki, M. & Nagasawa, H. (2007). The structure-function relationship analysis of Prismalin-14 from the prismatic layer of the Japanese pearl oyster, Pinctada fucata. *The FEBS Journal*, Vol. 274, No. 19, pp. 5158-5166, 1742-464X; 1742-464X

Takeuchi, T.; Sarashina, I.; Iijima, M. & Endo, K. (2008). In vitro regulation of CaCO(3) crystal polymorphism by the highly acidic molluscan shell protein Aspein. *FEBS Letters*, Vol. 582, No. 5, pp. 591-596, 0014-5793; 0014-5793

Testeniere, O.; Hecker, A.; Le Gurun, S.; Quennedey, B.; Graf, F. & Luquet, G. (2002). Characterization and spatiotemporal expression of orchestin, a gene encoding an ecdysone-inducible protein from a crustacean organic matrix. *The Biochemical Journal*, Vol. 361, No. Pt 2, pp. 327-335, 0264-6021; 0264-6021

Tohse, H.; Takagi, Y. & Nagasawa, H. (2008). Identification of a novel matrix protein contained in a protein aggregate associated with collagen in fish otoliths. *The FEBS Journal*, Vol. 275, No. 10, pp. 2512-2523, 1742-464X; 1742-464X

Tohse, H.; Saruwatari, K.; Kogure, T.; Nagasawa, H. & Takagi, Y. (2009). Control of Polymorphism and Morphology of Calcium Carbonate Crystals by a Matrix Protein Aggregate in Fish Otoliths. *Crystal Growth & Design*, Vol. 9, No. 11, pp. 4897-4901

Tompa, P. (2011). Unstructural biology coming of age. *Current Opinion in Structural Biology*, 1879-033X; 0959-440X

Tompa, P. (2005). The interplay between structure and function in intrinsically unstructured proteins. *FEBS Letters*, Vol. 579, No. 15, pp. 3346-3354, 0014-5793; 0014-5793

Tompa, P. (2002). Intrinsically unstructured proteins. *Trends in Biochemical Sciences*, Vol. 27, No. 10, pp. 527-533, 0968-0004; 0968-0004

Tsukamoto, D.; Sarashina, I. & Endo, K. (2004). Structure and expression of an unusually acidic matrix protein of pearl oyster shells. *Biochemical and Biophysical Research Communications*, Vol. 320, No. 4, pp. 1175-1180, 0006-291X; 0006-291X

Tye, C.E.; Hunter, G.K. & Goldberg, H.A. (2005). Identification of the type I collagen-binding domain of bone sialoprotein and characterization of the mechanism of interaction. *The Journal of Biological Chemistry*, Vol. 280, No. 14, pp. 13487-13492, 0021-9258; 0021-9258

Tye, C.E.; Rattray, K.R.; Warner, K.J.; Gordon, J.A.; Sodek, J.; Hunter, G.K. & Goldberg, H.A. (2003). Delineation of the hydroxyapatite-nucleating domains of bone sialoprotein. *The Journal of Biological Chemistry*, Vol. 278, No. 10, pp. 7949-7955, 0021-9258; 0021-9258

Uversky, V.N. (2009). Intrinsically disordered proteins and their environment: effects of strong denaturants, temperature, pH, counter ions, membranes, binding partners, osmolytes, and macromolecular crowding. *The Protein Journal*, Vol. 28, No. 7-8, pp. 305-325, 1875-8355; 1572-3887

Uversky, V.N. (2002). Natively unfolded proteins: a point where biology waits for physics. *Protein Science: A Publication of the Protein Society*, Vol. 11, No. 4, pp. 739-756, 0961-8368; 0961-8368

Uversky, V.N. & Dunker, A.K. (2010). Understanding protein non-folding. *Biochimica Et Biophysica Acta*, Vol. 1804, No. 6, pp. 1231-1264, 0006-3002; 0006-3002

Uversky, V.N.; Gillespie, J.R. & Fink, A.L. (2000). Why are "natively unfolded" proteins unstructured under physiologic conditions?. *Proteins*, Vol. 41, No. 3, pp. 415-427, 0887-3585; 0887-3585

Veis, A. (2003). Amelogenin gene splice products: potential signaling molecules. *Cellular and Molecular Life Sciences : CMLS*, Vol. 60, No. 1, pp. 38-55, 1420-682X; 1420-682X

Veis, A.; Sfeir, C. & Wu, C.B. (1997). Phosphorylation of the proteins of the extracellular matrix of mineralized tissues by casein kinase-like activity. *Critical Reviews in Oral Biology and Medicine: An Official Publication of the American Association of Oral Biologists*, Vol. 8, No. 4, pp. 360-379, 1045-4411; 1045-4411

Vymetal, J.; Slaby, I.; Spahr, A.; Vondrasek, J. & Lyngstadaas, S.P. (2008). Bioinformatic analysis and molecular modelling of human ameloblastin suggest a two-domain intrinsically unstructured calcium-binding protein. *European Journal of Oral Sciences*, Vol. 116, No. 2, pp. 124-134, 1600-0722; 0909-8836

Ward, J.J.; McGuffin, L.J.; Bryson, K.; Buxton, B.F. & Jones, D.T. (2004a). The DISOPRED server for the prediction of protein disorder. *Bioinformatics (Oxford, England)*, Vol. 20, No. 13, pp. 2138-2139, 1367-4803; 1367-4803

Ward, J.J.; Sodhi, J.S.; McGuffin, L.J.; Buxton, B.F. & Jones, D.T. (2004b). Prediction and functional analysis of native disorder in proteins from the three kingdoms of life. *Journal of Molecular Biology*, Vol. 337, No. 3, pp. 635-645, 0022-2836; 0022-2836

Wilt, F.H. (1999). Matrix and mineral in the sea urchin larval skeleton. *Journal of Structural Biology*, Vol. 126, No. 3, pp. 216-226, 1047-8477; 1047-8477

Wilt, F.H. & Ettensohn, C.A. (2007; 2008). The Morphogenesis and Biomineralization of the Sea Urchin Larval Skeleton, Wiley-VCH Verlag GmbH, 9783527619443

Wright, P.E. & Dyson, H.J. (1999). Intrinsically unstructured proteins: re-assessing the protein structure-function paradigm. *Journal of Molecular Biology*, Vol. 293, No. 2, pp. 321-331, 0022-2836; 0022-2836

Wu, H.; Teng, P.N.; Jayaraman, T.; Onishi, S.; Li, J.; Bannon, L.; Huang, H.; Close, J. & Sfeir, C. (2011). Dentin matrix protein 1 (DMP1) signals via cell surface integrin. *The Journal of Biological Chemistry*, 1083-351X; 0021-9258

Wustman, B.A.; Morse, D.E. & Evans, J.S. (2004). Structural characterization of the N-terminal mineral modification domains from the molluscan crystal-modulating biomineralization proteins, AP7 and AP24. *Biopolymers*, Vol. 74, No. 5, pp. 363-376, 0006-3525; 0006-3525

Wustman, B.A.; Santos, R.; Zhang, B. & Evans, J.S. (2002). Identification of a "glycine-loop"-like coiled structure in the 34 AA Pro,Gly,Met repeat domain of the biomineral-associated protein, PM27. *Biopolymers*, Vol. 65, No. 5, pp. 362-372, 0006-3525; 0006-3525

Wustman, B.A.; Weaver, J.C.; Morse, D.E. & Evans, J.S. (2003). Structure-function studies of the lustrin A polyelectrolyte domains, RKSY and D4. *Connective Tissue Research*, Vol. 44 Suppl 1, pp. 10-15, 0300-8207; 0300-8207

Wuttke, M.; Muller, S.; Nitsche, D.P.; Paulsson, M.; Hanisch, F.G. & Maurer, P. (2001). Structural characterization of human recombinant and bone-derived bone sialoprotein. Functional implications for cell attachment and hydroxyapatite binding. *The Journal of Biological Chemistry*, Vol. 276, No. 39, pp. 36839-36848, 0021-9258; 0021-9258

Xu, G. & Evans, J.S. (1999). Model peptide studies of sequence repeats derived from the intracrystalline biomineralization protein, SM50. I. GVGGR and GMGGQ repeats. *Biopolymers*, Vol. 49, No. 4, pp. 303-312, 0006-3525; 0006-3525

Zhang, B.; Wustman, B.A.; Morse, D. & Evans, J.S. (2002). Model peptide studies of sequence regions in the elastomeric biomineralization protein, Lustrin A. I. The C-domain consensus-PG-, -NVNCT-motif. *Biopolymers*, Vol. 63, No. 6, pp. 358-369, 0006-3525; 0006-3525

Zhang, B.; Xu, G. & Evans, J.S. (2000). Model peptide studies of sequence repeats derived from the intracrystalline biomineralization protein, SM50. II. Pro,Asn-rich tandem repeats. *Biopolymers*, Vol. 54, No. 6, pp. 464-475, 0006-3525; 0006-3525

Zou, Y.; Wang, H.; Shapiro, J.L.; Okamoto, C.T.; Brookes, S.J.; Lyngstadaas, S.P.; Snead, M.L. & Paine, M.L. (2007). Determination of protein regions responsible for interactions of amelogenin with CD63 and LAMP1. *The Biochemical Journal*, Vol. 408, No. 3, pp. 347-354, 1470-8728; 0264-6021

2

Single Amino Acids as Additives Modulating $CaCO_3$ Mineralization

Christoph Briegel and Jong Seto
Department of Chemistry, University of Konstanz, Konstanz
Germany

1. Introduction

Biomineralization has become an important theme in the fields of biology, chemistry, and materials science. It is an ability of organisms to produce organic-inorganic composite structures, which mainly serve a function for storage, protection or skeletal support. What makes biomineralization such an interesting subject of research is the simplicity organisms are able to modify their surroundings into complex materials. If you simply compare biogenic and geologically produced minerals with the same chemical compositions, you will notice a substantial difference in ultrastructure and subsequently, difference in properties.

There are many different examples where biomineralization can be witnessed, like in sea coral, teeth or eggshells. The diversity of structures, minerals and macromolecules that build up these tissues is impressive (1, 2, 3). Diatoms are able to precipitate silica from the environment to create their ornate skeletons. Another example of biomineralization is the mineral magnetite with its ferromagnetic properties. Not only it is found in many microorganisms as a structural component, but it can furthermore be used for navigation along the Earth's magnetic field. In all these diverse examples of biomineralization processes, a common theme can be found in each case—mineralization occurs in the presence of an organic matrix that appears to direct mineral and ultrastructural morphology (4-6) (7) (8) (9-11) (12). A closer look at the mineralized tissues of certain invertebrates shows that $CaCO_3$ is the predominant mineral composition utilized, but have completely different structural features from its geological counterparts. It is this same geological mineral, which is usually brittle and very unstable under shear forces, that provides strength and structure to invertebrates. The combination of organic and inorganic materials and the control of growth make these biologically mineralized structures mechanically tough and highly resilient materials (13).

In Nature, $CaCO_3$ can be found in three different forms of mineral-modifications (14). The most frequent and thermodynamically stable form is calcite. It consists of trigonal crystals and appears mostly transparent or milky in solution. Also well-known are vaterite and aragonite. They both have the same chemical stoichiometry but form a crystalline lattice that is different from calcite due to the differences in packing of the unit cells in each respective polymorph. Aragonite is metastable and slowly transforms to calcite after some time.

Aragonitic crystals have a prismatic structure. Vaterite is also metastable and is rare, but when it does form, it occurs in orthorhombic unit cells (Figure 1).

A significant form of $CaCO_3$ and until recently, found to be as ubiquitious as its crystalline forms, emerges in the first stage of pre-nucleation, is amorphous calcium carbonate (ACC). Addadi and coworkers found that an amorphous phase, actually composed of several independent phases, was involved in the formation of certain biomineralized structures (15) (16). Gilbert et al. reports the thermodynamics of ACC I and ACC II types during transformation to crystalline products (17). Gebauer and coworkers demonstrated that these two different species of ACC control the subsequent formation of calcite or vaterite. Furthermore, they showed that a part of Ca^{2+} ions forms neutral equilibrium clusters before nucleation of the supersaturated solution arises (18). Several groups have also investigated the influence of different additives on the nucleation of $CaCO_3$ (19-22). Amongst others, polycarboxylates which usually serve as inhibitors of the formation of scale in laundry detergents or dishwaters were used to investigation inhibition of nucleation (23). The experiments revealed that additives and acidic polymers have very different effects on nucleation such as adsorption of calcium ions or an influence on soluble-cluster formation (18). These previous studies show that nucleation of calcium carbonate is a complex, multistep process whereby additives often are a major determinant driving crystallization of $CaCO_3$ (18) (24).

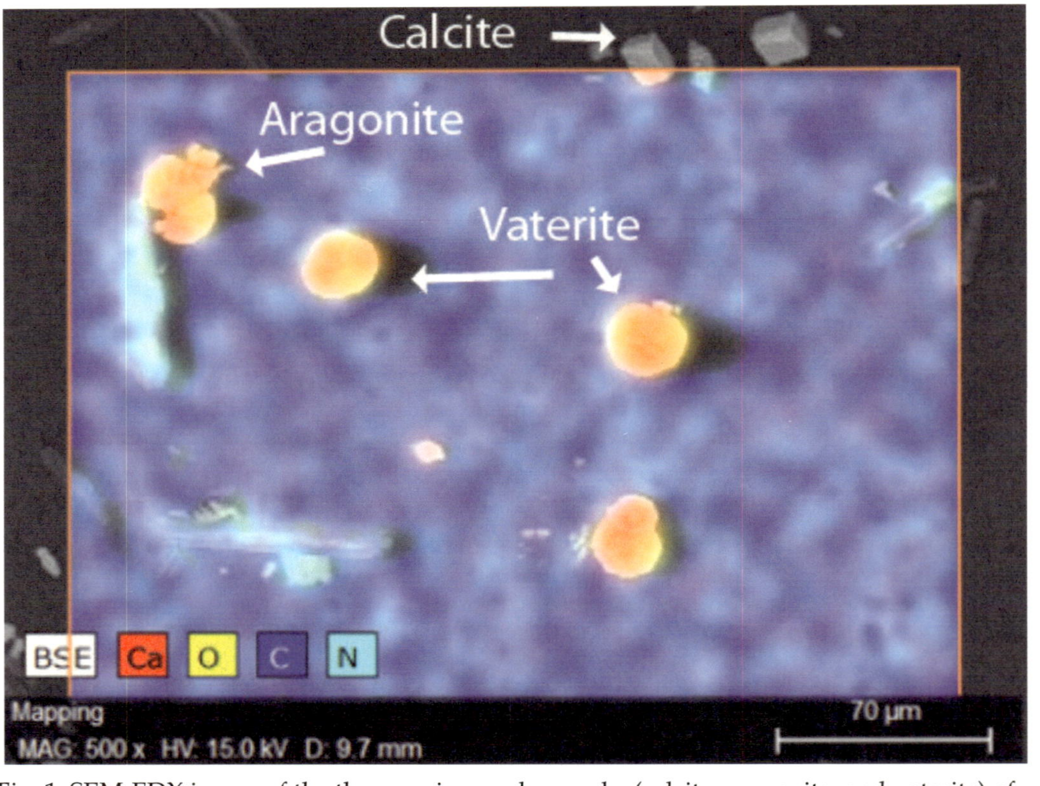

Fig. 1. SEM-EDX image of the three various polymorphs (calcite, aragonite, and vaterite) of $CaCO_3$ mineral

2. Amino acids

Several groups have examined the effect of biological additives, proteins in their native conformations, on the effects of calcium carbonate mineralization. From secondary to tertiary conformations, several aspects of structure modulate the effect--including the magnitude of effect specific proteins have on crystallization. To simplify the understanding of these protein interactions, we attempt to describe the influence of their constituent amino acid groups. In this manner, structure is removed in order to determine the specific effects of amino acids on crystallization. There are twenty so called natural amino acids, which differ in their side chain functional groups (25). Eight of them are considered to be essential that means that they cannot be produced from other compounds by the human body (25). Some of these relevant amino acids are found to modulate calcium carbonate mineralization. Their characteristics are briefly described in the following sections.

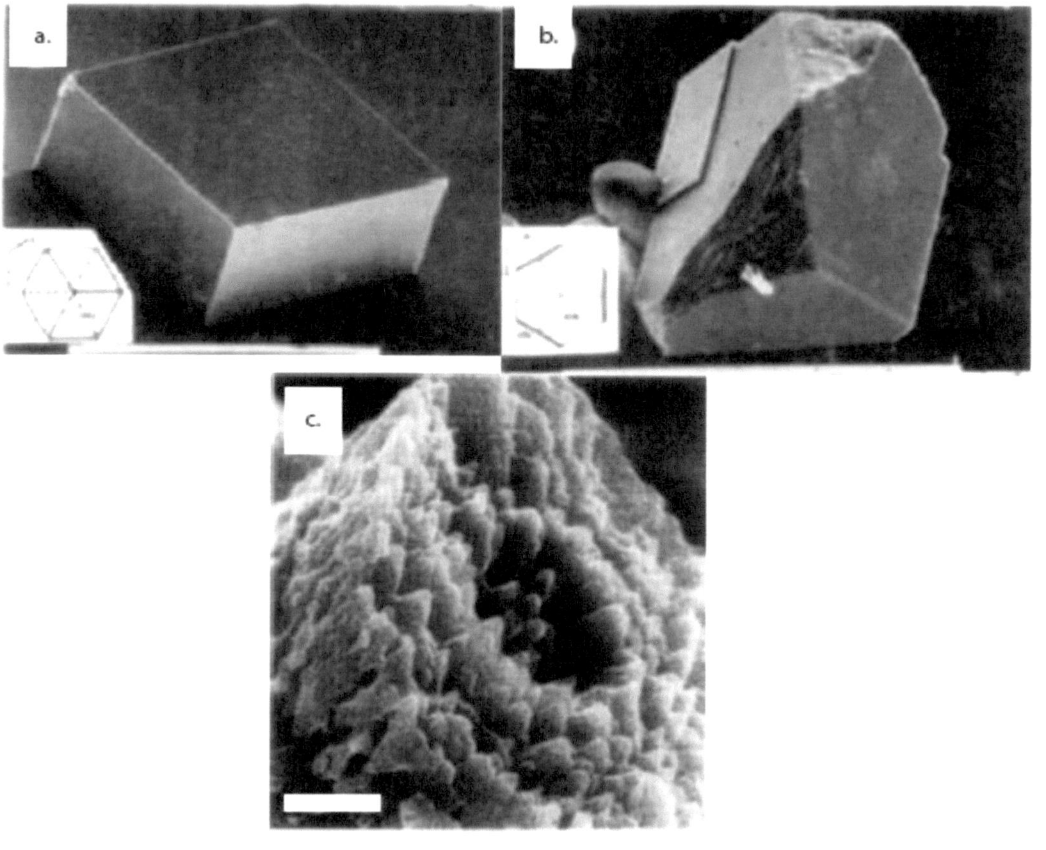

Fig. 2. Various morphologies of $CaCO_3$ single crystalline precipitates (a.) geological calcite (b.) calcite grown in the presence of sea urchin spicule matrix proteins [scale bar = 100 μm] (c.) fracture surface of a broken sea urchin spicule [scale bar = 500 nm]

Arginine is a basic amino acid with a pK_a of 12.48. The side chain contains a complex guanidinium group which is positively charged under neutral, acidic and even most basic conditions. This explains the alkaline characteristics. Asparagine is a polar amino acid with a carboxamide group in its side chain. Asparagine as well as arginine are nonessential amino acids. Glutamic acid belongs to the acidic amino acids with a pK_a of 4.1. Its side chain contains a carboxylic acid at the end. Furthermore, glutamic acid does not belong to the essential amino acids. In contrast, methionine is one of the few essential amino acids. Together with cysteine, these amino acids are the only two whose side chain consists of a sulphur group. Proline has a unique structure among the twenty amino acids because of its secondary α-amino group. The cyclic structure of proline shows nonpolar behavior. Serine is a polar amino acid due to the hydroxyl group in the side chain. Valine is an essential amino acid. Together with leucine and isoleucine, it belongs to the branched-chain amino acids. Due to its alkyl side chain, it shows a nonpolar character (25).

However, it is not yet known which amino acids are responsible for the different tasks concerning crystal nucleation and formation, especially their functional side groups. From this these questions about function and structural characteristics, it is particularly interesting to investigate the effects of crystallization of $CaCO_3$ with the addition of different amino acids as additives. Crystal morphology as well and crystallization kinetics in $CaCO_3$ formation have been specifically examined in the presence of specific amino acid sequences (Figure 2).

All twenty amino acids have been surveyed and the amino acids that had dramatic morphology effects on the $CaCO_3$ precipitates in comparison to their geological analogs were further characterized with other methods.

3. Vapor-diffusion crystallization

The precipitation of calcium carbonate offers a model system for the investigation of nucleation and subsequent crystal growth (26). A simple, but still effective method to monitor various precipitations is the "vapor diffusion technique." In the case of $CaCO_3$, it is based on the decomposition of $(NH_4)_2CO_3$ or NH_4HCO_3 into CO_2 and NH_3 (Figure 3). The precipitation of a solid phase is described by the following equations:

$$CO_2 + H_2O \leftrightarrow HCO_3^- + H^+ \qquad (1)$$

$$Ca^{2+} + HCO_3^- \leftrightarrow CaCO_3 + H^+ \qquad (2)$$

whereby Equation (1) describes the formation of bicarbonate ions from CO_2 and water as an initial step to formation of calcium carbonate in Equation (2). The reaction is performed in a desiccator which provides an isolated environmental chamber. In our case, the "vapor-diffusion crystallization" technique, is a variation of "vapor diffusion," has been used. For this purpose, a dish of NH_4HCO_3 solution was located at the bottom of the chamber. Drops of $CaCl_2$ solution were placed on cover slides, which were located on a shelf above the solution (Figure 3). Despite its easy handling, this method has also disadvantages. Disadvantages include low reproducibility of the precipitation process and difficulties in monitoring (26). Another disadvantage is the dependency of the precipitation on the volume of solution and desiccator (26).

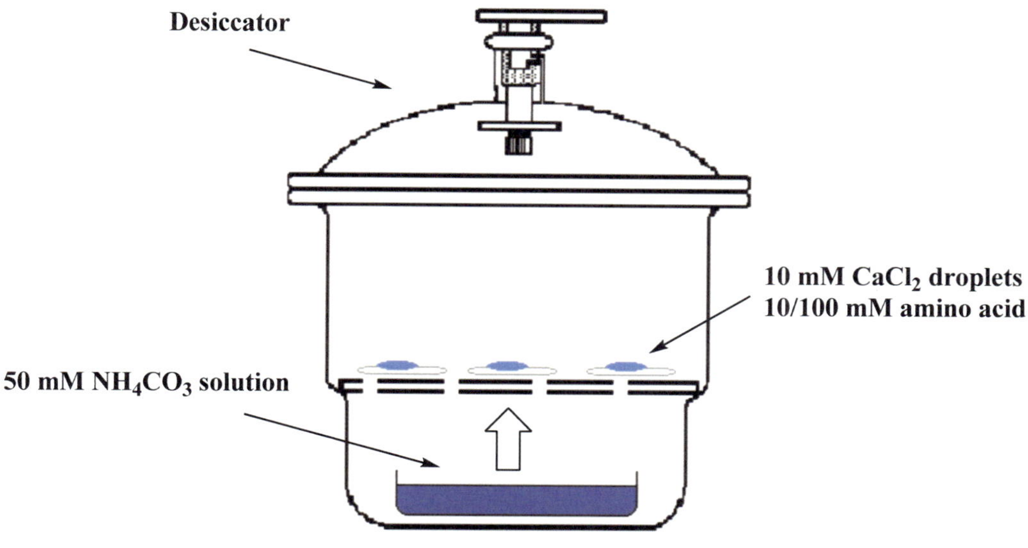

Fig. 3. Diagram of the vapor diffusion sitting drop crystallization experiment. At the bottom of the desiccator is a dish with 50 mM NH_4CO_3 solution. On a shelf above are placed cover slides with 10 mM $CaCl_2$ solution.

4. *In situ* potentiometric titration

To observe the kinetics of $CaCO_3$ crystallization, from ion clusters to post-nucleation aggregates, *in situ* potentiometric titration has been utilized. Specifically, the precipitation of $CaCO_3$ mineral products in the presence of amino acids in solution can be observed and compared to reference situations.

The basis of this measurement is based on the potential differences across a membrane and deriving from this a quantity that can be correlated to ion concentrations. By utilizing the Nernst equation, a potential to determine the amount of free Ca^{2+} ions and bound Ca^{2+} ions in solution can be determined as follows:

$$E = E_0 + \frac{RT}{nF} ln\left(\frac{[free\ Ca^{2+}\ ions]}{[bound\ Ca^{2+}\ ions]}\right)$$

where E_0 is the starting chemical potential, R is the gas constant, F is Faraday's constant, and T is temperature.

The other quantities of the equation are measured by a commercial, computer controlled titration system equipped with pH and Ca^{2+} selective electrodes to measure the respective quantities in solution as $CaCl_2$ is titrated into the reaction solution (Figure 4). The quantity can then be correlated to the concentration of Ca^{2+} ions in solution and as well concentration bound Ca^{2+} ions.

In our studies, the titrations were carried out in a 10 mM carbonate buffer. The carbonate buffer contained a mixture of Na_2CO_3 and $NaHCO_3$. The pH-value was set to 9.75. For precipitation, a 10 mM $CaCl_2$ solution was used. The following amino acids were used as additives: asparagine, arginine, glutamic acid, methionine, proline, serine and valine. Their concentrations were 10 mM and 100 mM, respectively. All titrations were performed at room temperature. The experiment was carried out in a 100 ml beaker filled with 10 ml of 10 mM carbonate buffer accomplished with 10 mM/100 mM amino acid. Before every titration, constant pH values were assured using 100 mM sodium hydroxide and 10 mM hydrochloride acid. The application rate was 5 µl for NaOH/ HCl and 10 µl for $CaCl_2$ solution. For the titration of 100 mM asparagine, glutamic acid and arginine required more alkaline buffer solution was required. For the titration with glutamic acid, a $NaCO_3$/ NaOH buffer was used. Its pH-value was set to 10.4. In case of asparagine and arginine, the pH-value of $NaCO_3$ buffer was set by manually adding 1 M NaOH to get a pH of 9.75. All titrations were carried until reaching mineral precipitation.

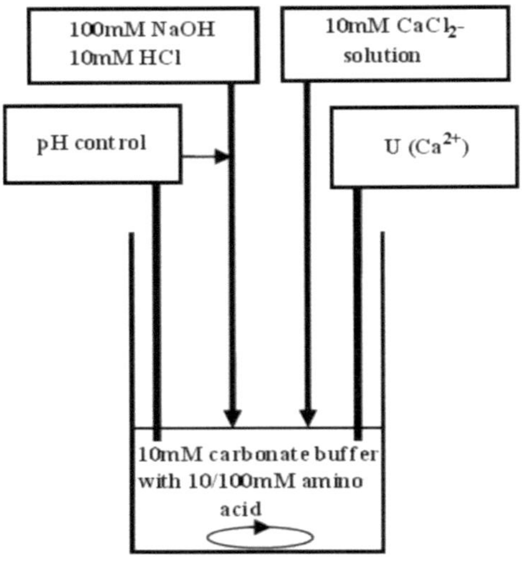

Fig. 4. Scheme of the titration experiment. 10mM $CaCl_2$-solution was titrated into a beaker containing 10mM carbonate buffer with 10mM/100 mM concentrations of amino acid. For constant pH, 100 mM NaOH/ 10 mM HCl were added, respectively. The titration was monitored with a Ca^{2+} ion electrode.

5. Discussion

The effects of specific amino acids on the crystallization of $CaCO_3$ were investigated. For this purpose, "vapor-diffusion crystallization" measurements were used to initally survey the amino acids having an effect on crystal morphologies. All twenty natural amino acids were surveyed and the resulting crystals were characterized by various techniques such as light microscopy and scanning electron microscopy (Table 1,2). In the most interesting cases of an amino acid, further investigations were used to characterize the specific amino acids. Subsequently, *in situ* potentiometric measurements were conducted to determine the differential kinetics involved in each of the amino acid cases.

	Amino acid	Properties	Average size of crystals
1.	Reference ($CaCl_2$ only)	Accumulation of undefined crystals; few calcite structures	13 μm (only calcite structure)
2.	Alanine	Some calcite crystals; few round undefined structures; widely scattered	4 μm
3.	Aspartic acid	Accumulation of calcite-like crystals; smooth borders	10 μm
4.	Cysteine	LM: Accumulation of vaterite-like clusters SEM: only calcite-like-clusters (very small amount)	9 μm
5.	Glutamine	Calcite crystals	14 μm
6.	Glycine	Accumulation of vaterite-like forms; a few calcite clusters	17 μm
7.	Histidine	Vaterite and calcite structures	23 μm
8.	Isoleucine	Calcite structure; widely scattered	17 μm
9.	Leucine	Calcite structures; few vaterite-crystals; widely scattered	16 μm
10.	Lysine	Small crystals; calcite structures; chains of crystals	6 μm
11.	Phenylalanine	Vaterite/calcite structures; widely scattered	16 μm
12.	Threonine	Accumulation of calcite crystals; sharp borders	11 μm
13.	Tryptophan	Vaterite/calcite structures; accumulation of clusters	12 μm
14.	Tyrosine	Vaterite/calcite structures	17 μm

Table 1. Description of crystalline precipitates mineralized in the presence of amino acids (10 mM) in 10 mM $CaCl_2$-solution as observed in polarized light microscopy(V: 15 μl), duration: 72 hours

Two series with a 10 mM and the other with 100 mM amino acid concentration were carried out as an approximate "low" and "high" concentration, respectively (Table 1, 2). It could be observed that, in general, in both cases the calcium carbonate crystals possess the similar size dimensions (between 10 and 50 μm) and show calcite-like forms (Figure 5, 6). Nevertheless, several differences can be recognized, particularly for the seven selected samples. Here, more vaterite-like structures than calcite in the samples with a lower amino acid concentration can be observed. Moreover, the size of the formed $CaCO_3$ crystals were rather smaller (between 5 -30 μm). It can be observed that amino acids have a distinct effect on the crystallization of $CaCO_3$ (Figure 5,6). In contrast, the reference situations (without additives), only small trigonal crystals with sharp edges were found. The $CaCO_3$ samples with amino acid show usually rounded crystals with smooth edges and partially entirely new structures. The amino acids arginine, asparagine, glutamic acid, proline, methionine, serine, valine were subsequently determined to produce dramatic effects in crystallization and chosen for further characterization (27).

	Amino acid	Properties	Average size of crystals
1.	Reference ($CaCl_2$ only)	Sharp borders; widely scattered; only calcite structures	31 μm
2.	Alanine	Sharp borders; widely scattered; only calcite structures	16 μm
3.	Aspartic acid	Branched undefined structures	Not measurable
4.	Cysteine	Accumulations of round vaterite-like clusters; some calcite crystals	Not measurable
5.	Glutamine	Calcite structures; accumulation of clusters	15 μm
6.	Glycine	Smooth borders; widely scattered; calcite form	11 μm
7.	Histidine	Smooth borders; widely scattered; calcite form	Not measurable
8.	Isoleucine	Calcite structures	11 μm
9.	Leucine	Some calcite crystals; few undefined structures	27 μm
10.	Lysine	Small accumulations of crystals; calcite-like forms; smooth borders	Not measurable
11.	Phenylalanine	Small crystals; calcite-like structures	9 μm
12.	Threonine	Calcite-like structures; widely scattered between amino acid	50 μm
13.	Tryptophan	Calcite structures; widely scattered between amino acid	Different sizes
14.	Tyrosine	A few calcite structures; crystals covered by amino acid	Not measureable

Table 2. Description of crystalline precipitates mineralized in the presence of amino acids (100 mM) in 10 mM $CaCl_2$-solution. (V: 15 μl), duration: 18 hours observed in polarized light microscopy

In the "*in situ* titration" experiment, the exact phases (saturation, supersaturation, nucleation and growth) of calcium carbonate crystallization in the presence of amino acids can be monitored (Figure 7). In this experiment the titration was also performed with 10 mM and 100 mM amino acid concentration. Thereby, a 10 mM $CaCl_2$ solution was titrated in a beaker with 10 mM carbonate buffer with an amino acid. The titration was monitored with a Ca^{2+}- and pH-electrode. During titration the pH-value was kept constant to exclude pH effects on the calcium carbonate crystallization. There were problems with the samples 100 mM arginine, asparagine and glutamic acid. In these cases, the amino acids were too acidic, so that the standard buffer did not perform to keep the pH of the solution constant. For glutamic acid, a new buffer consisting of 10 mM sodium carbonate and 1 M sodium hydroxide, was used. For the amino acids asparagine and arginine, the buffer solution was set up manually before titration. In this 1 M NaOH was added to the carbonate buffer until pH achieved 10. It should be noted that in these cases, the volume of the carbonate buffer was not 10 ml anymore. It is hence difficult to compare these titrations with the other samples due to different volumes.

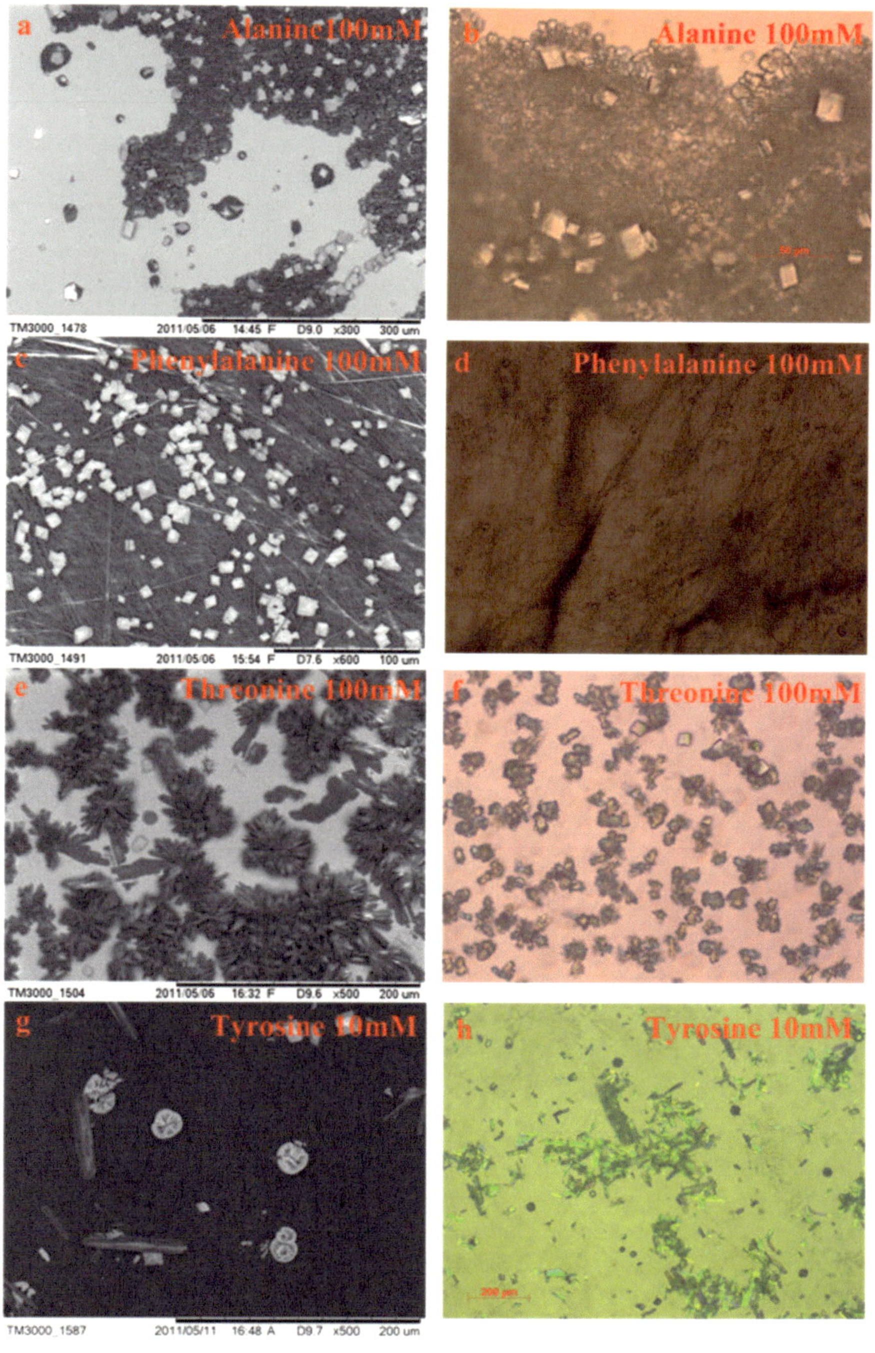

Fig. 5. Light microscopy and SEM pictures of crystalline precipitates after $CaCO_3$ vapor diffusion crystallization. The left column shows SEM recordings. The right one light microscope images, accordingly.

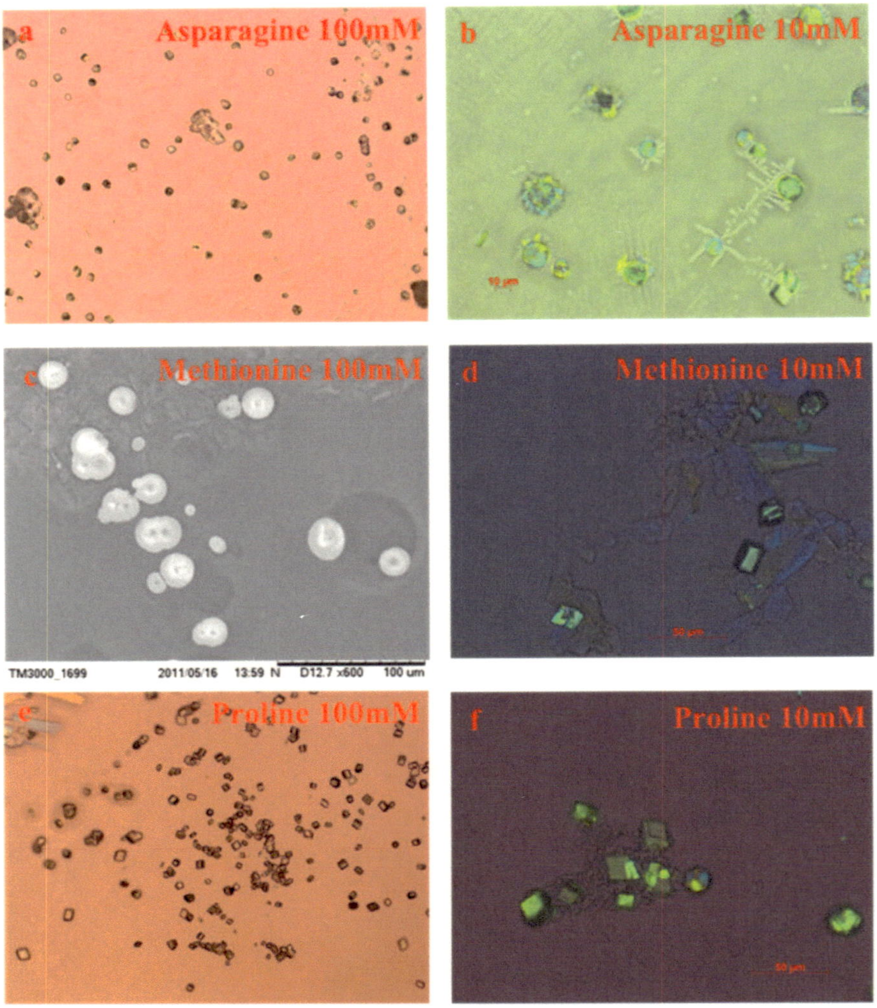

Fig. 6. Light microscopy and SEM images of $CaCO_3$ crystals after vapor-diffusion crystallization in the presence of asparagine, methionine or proline.

All titration curves show the typical features of a Lamer diagram. The greatest effect on the nucleation in the 10 mM experiment was found to be glutamic acid (Figure 7). In comparison to the reference, the free calcium concentration was four times larger before nucleation started. It is surprising that the samples proline and arginine have even a negative effect because the nucleation $CaCO_3$ started earlier than in the reference. It can be possible that the interaction of Ca^{2+} ions with the amino acid leads to a lower activation barrier and thus to an earlier nucleation stage. In the second titration assay with 100 mM, valine showed the biggest effect on the nucleation of calcium carbonate. In this case, an obviously higher concentration of free Ca^{2+} ions as in the reference was measured. Nevertheless, the comparison of both graphs (10 mM /100 mM) shows some interesting observations. By increasing the amino acid concentration ten times higher,a doubling of the critical concentration of calcium ion can be observed and the nucleation also took twice as long. Consequently, it can be said that the interaction of amino acid molecules and Ca^{2+} ions lead to a delay of nucleation and further to a change in crystallization conditions. Since in this these titration measurements (10 mM/100 mM) at two concentrations, these results have to be dealt critically. Subsequent measurements should provide certainty on the exact interactions regulating the crystallization processes.

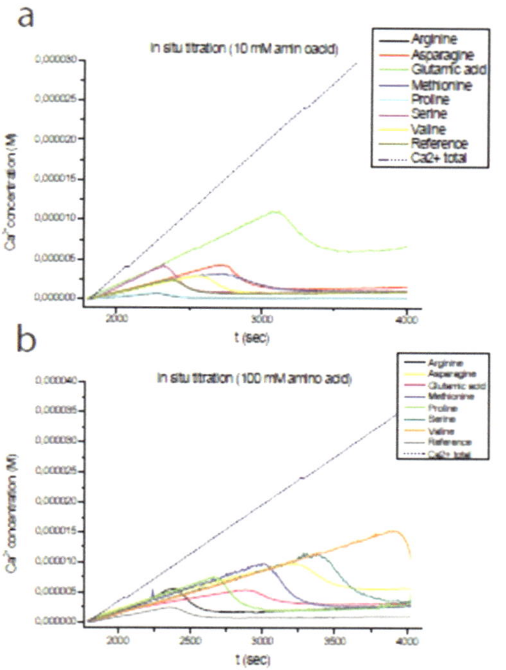

Fig. 7. Titration curves of calcium carbonate precipitation in the presence of a) 10 mM and b) 100 mM concentration amino acids

Light microscope measurements showed that the formed crystals were much smaller in the titration experiment than the "vapor-diffusion crystallization" method. Since it was difficult to observe crystals or different structures at all with light microscopy due to the size of the crystals, further SEM and TEM images were taken. With SEM, only scattered agglomerations of $CaCO_3$ were found, but there was a significant difference among the samples. All samples of 100 mM amino acid concentrations showed round forms in contrast to the dilution experiment. Here, only clusters of calcite-like structures were found, which have rounded edges and smooth corners. Perhaps higher concentrations of amino acids can promote thermodynamically unfavored amorphous phases of calcium carbonate. In diluted samples, further TEM measurements show only round structures which developed into big agglomerations (Figure 8, 9). Exceptions were found in the samples arginine and asparagine. In these samples, also calcite-like forms could be found. Comparison of SEM and TEM data (titration experiment 10 mM amino acid concentration), show no definitive conclusions between the two measurements. At a first glance, the preparation of the samples for SEM and TEM are different and thus provide different results. The more important point is that each measurement type operates at different lengthscale. The SEM pictures are in a size range of few micrometers, whereas the TEM in contrast operates at the nanometer range and additionally the samples have to be very thin (few nanometers) to see visualize. The very small round crystals and clusters that are seen on the TEM images are not detectable with SEM or LM. The much bigger crystals, which later usually develop into calcite-like structures, can be observed with light microscopy or SEM.

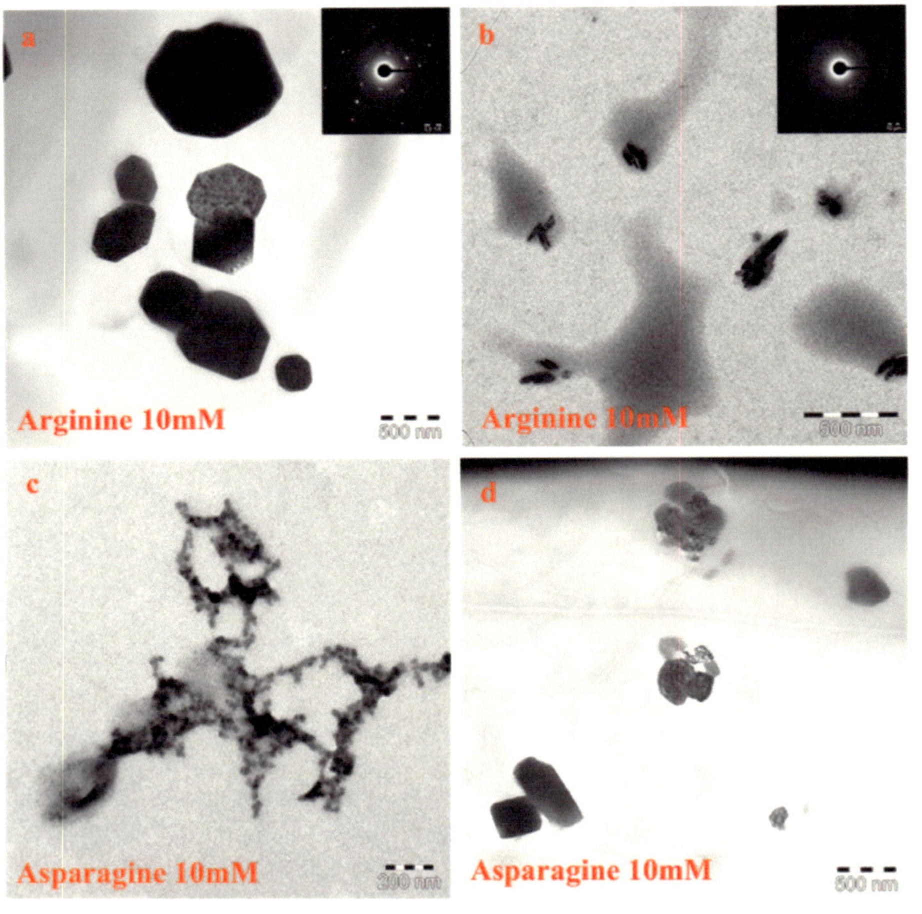

Fig. 8. TEM from the *in situ* potentiometric titration measurements with 10 mM amino acid concentration a & b) arginine, c & d) asparagine.

It is assumed that the different stages of crystal growth of calcium carbonate are measured in each of the different measurements under LM, SEM, and TEM. After nucleation, there are the two types of ACC I or ACC II. This could be the small round structures, which can be seen on the TEM pictures. After some time, more and more round crystals agglomerate to the bigger cluster as seen in Figure 9d and 9g, for instance.

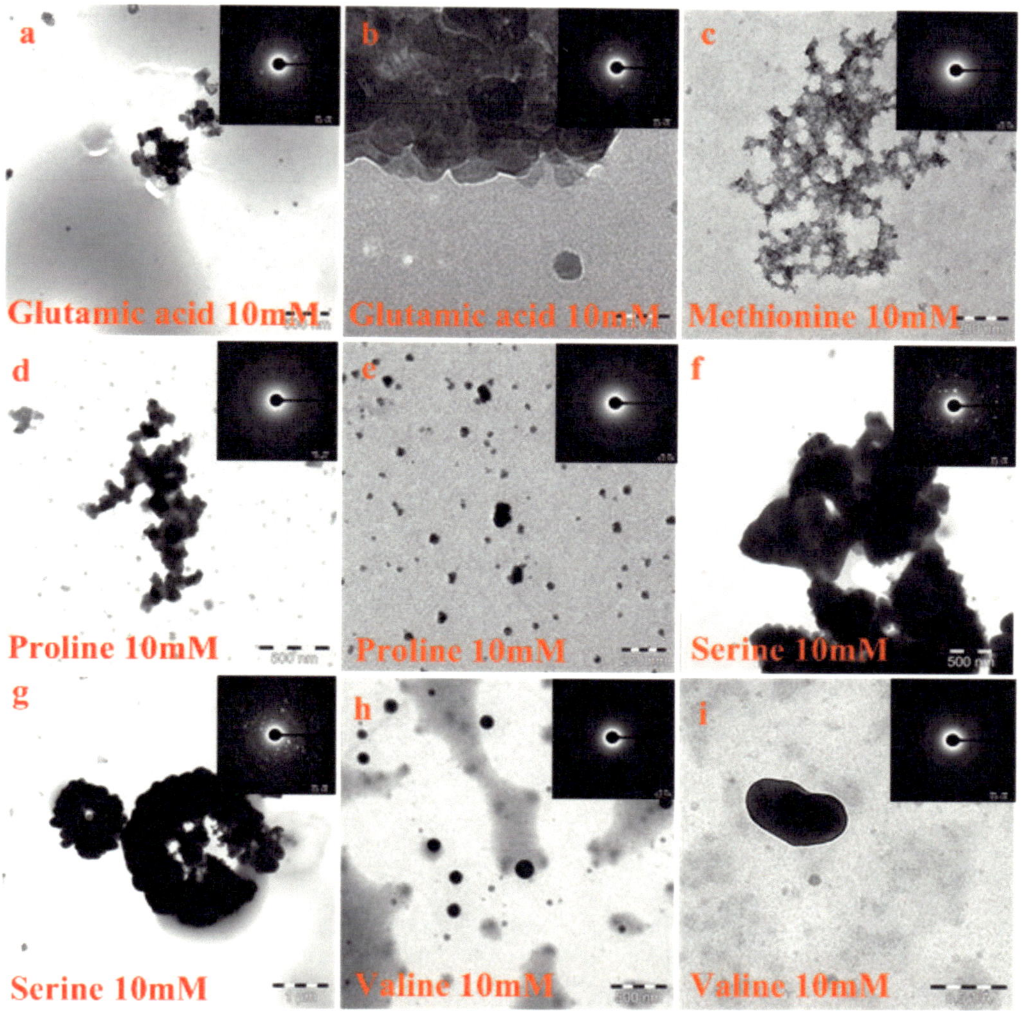

Fig. 9. TEM from *in situ* potentiometric titration measurements with 10 mM amino acid concentration a & b) glutamic acid, c) methionine, d & e) proline, f & g) serine and h & i) valine.

Finally, the instable clusters convert to a thermodynamically stable crystal structure, like calcite or vaterite, which can be seen on the TEM pictures (Figure 10). The amino acids can serve as nucleation- and growth- promoting molecules for calcium carbonate, reducing the activation energy of nucleation and facilitating the crystal growth. Interactions at these early stages of nucleation can prefer or inhibit specific crystal shapes and hence, control structures during calcification. In the "in-situ-titration," much smaller crystals can be produced. The "vapor-diffusion crystallization" experiment provides $CaCO_3$ more points for heterogeneous nucleation than "in-situ-titration." This lowers the activation barrier and thus, results in faster growth and finally larger crystals. In conclusion, it was shown that specific amino acids have a distinct effect on calcification. Depending on the type of amino acid and concentration, calcium carbonate can form different structures. Round structures, like vaterite are often favored instead of the thermodynamically stable calcite as intermediate structures. Furthermore, the type of crystallization also plays a decisive role as shown here in the aforementioned two different crystallization situations.

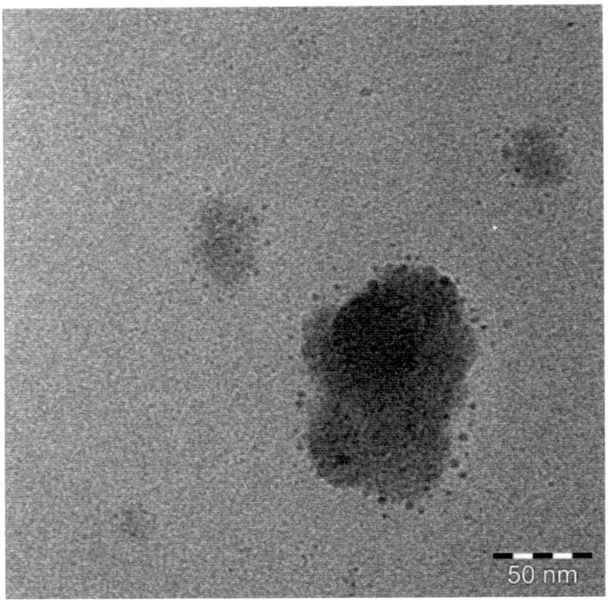

Fig. 10. TEM of a titration experiment with 10 mM of proline. A $CaCO_3$ crystal surrounded of many small crystals is shown.

6. Outlook

Use of *in situ* solution based techniques to study mineralization in the presence of additives is a powerful tool to determine kinetic as well as thermodynamic parameters involved in driving mineralization processes. Specifically, with the ability to determine the binding kinetics of pre-nucleation clusters, amorphous aggregates, or inhibition kinetics, Nature's use of additives in regulating mineral growth and development can be closely followed as it occurs. With the addition of other *in situ* physical measurement methods (i.e. *in situ* AFM), the ability to study the influence of additives, both small molecules as well as supramolecular complexes, can be accomplished with better observations of key events like nucleation or ripening stages in mineralization. By fully elucidating these processes in biomineralization, a better understanding on composing stronger tougher materials is achieved such that synthetic materials can be tailored with similar materials and mechanical properties found in Nature.

7. References

[1] Weiner S (2008) Biomineralization: A structural perspective. *J. Struct. Bio.* 163:229-234.
[2] Weiner S, Addadi, L. (1997) Design strategies in mineralized biological materials. *Journal of Materials Chemistry* 7(5):689-702.

[3] Lowenstam HA, Weiner, S. (1989) *On Biomineralization* (Oxford University Press, New York) p 323.
[4] Fang PA, Conway JF, Margolis HC, Simmer JP, & Beniash E (2011) Hierarchical self-assembly of amelogenin and the regulation of biomineralization at the nanoscale. *PNAS* 108(34).
[5] Berman A, Addadi, L., Kvick, A., Leiserowitz, L., Nelson, M., Weiner, S. (1990) Intercalation of sea-urchin proteins in calcite- Study of a cyrstalline composite-material. *Science* 250(4981):664-667.
[6] Berman A, Addadi L, & Weiner S (1988) Interactions of sea urchin skeleton macromolecules with growing calcite crystals--a study of intracrystalline proteins. *Nature* 331:546-548.
[7] Seto J, Zhang, Y., Hamilton, P., Wilt, F. (2004) The localization of occluded matrix proteins in calcareous spicules of sea urchin larvae. *Journal of Structural Biology* 148(1):123-130.
[8] Veis A, Barss, J., Dahl, T., Rahima, M., Stock, S. (2002) Mineral-Related Proteins of Sea Urchin Teet: Lytechinus variegates. *Microscopy Research and Technique* 59:342-351.
[9] Wilt FH (1999) Matrix and Mineral in the Sea Urchin Larval Skeleton. *Journal of Structural Biology* 126:216-226.
[10] Suzuki M, *et al.* (2009) An Acidic Matrix Protein, Pif, is a Key Macromolecule for Nacre Formation. *Science* 325(5946):1388-1390.
[11] Kroeger N, Lorenz S, Brunner E, & Summper M (2002) Self-assembly of highly phosphorylated silaffins and their function in biosilica morphogenesis. *Science* 298:584-586.
[12] Levy-Lior A, Weiner S, & Addadi L (2003) Achiral calcium-oxalate crystals with chiral morphology from the leaves of some Solanacea plants. *Helvetica Chimica Acta* 86(12):4007-4017.
[13] Cusack M & Freer A (2008) Biomineralization: Elemental and Organic Influence in Carbonate Systems. *Chem. Rev.* 108(11):4433-4454.
[14] Anthony JW, Bideaux RA, Bladh KW, & Nichols MC eds (2001) *Handbook of Mineralogy* (Mineralogical Society of America, Chantilly, VA), Vol 5.
[15] Raz S, Hamilton P, Wilt FH, Weiner S, & Addadi L (2003) The transient phase of amorphous calcium carbonate in sea urchin larval spicules: the involvement of proteins and magnesium ions in its formation and stabilization. *Adv. Func. Mat.* 13(6):480-486.
[16] Addadi L, Raz S, & Weiner S (2003) Taking Advantage of Disorder: Amorphous Calcium Carbonate and Its Roles in Biomineralization. *Adv. Mat.* 15(12):959-970.
[17] Radha AV, Forbes TZ, Killian C, Gilbert PUPA, & Navrotsky A (2010) Transformation and crystallization energetics of synthetic and biogenic amorphous calxium carbonate. *PNAS* 107(38):16438-16443.
[18] Gebauer D, Coelfen H, Verch A, & Antonietti M (2008) The Multiple Roles of Additives in CaCO3 Crystallization: A Quantitative Case Study. *Adv. Mat.* 21(4):435-439.
[19] Amos FFE, J.S. (2009) AP7, a Partially Disordered Pseudo C-Ring Protein, Is Capable of Forming Stabilized Aragonite in Vitro. *Biochemistry* 48:1332-1339.
[20] Xie AJ, Shen YH, & Zhang CY (2005) Crystal growth of calcium carbonate with various morphologies in different amino acid systems. *Journal of Crystal Growth* 285(3):436-443.

[21] Ndao M, *et al.* (2010) Intrinsically Disordered Mollusk Shell Prismatic Protein That Modulates Calcium Carbonate Crystal Growth. *Biomacromolecules* 11:2539-2544.

[22] Elhadj S, *et al.* (2006) Peptide Controls on Calcite Mineralization: Polyaspartate Chain Length Affects Growth Kinetics and Acts as a Stereochemical Switch on Morphology. *Crystal Growth and Design* 6(1):197-201.

[23] Addadi L, Moradian-Oldak J, & Weiner S (1991) Macromolecule-Crystal Recognition in Biomineralization: Studies using synthetic polycarboxylate analogs. in *Surface Reactive Peptides and Polymers*, eds Sikes CS & Wheeler AP (ACS Symposium Series, Dallas, Texas), pp 13-27.

[24] Gebauer D, Voelkel A, & Coelfen H (2008) Stable Prenucleation Calcium Carbonate Clusters. *Science* 322:1819-1822.

[25] Lehninger AL (2008) *Principles of Biochemistry* (W.H. Freeman) 5th Ed.

[26] Gomez-Morales J, Hernandez-Hernandez A, Sazaki G, & Garcia-Ruiz JM (2009) Nucleation and polymorphism of calcium carbonate by a vapor diffusion sittig drop crystallization technique. *Crystal Growth and Design* 10(2):963-969.

[27] Seto J, Briegel C, & Coelfen H (2012) Surveying the effects of individual amino acids on CaCO3 mineralization. in *Zeitschrift fuer Kristallographie* (Oldenbourg Wissenschaftsverlag GmbH, Muenchen).

3

Control of $CaCO_3$ Crystal Growth by the Acidic Proteinaceous Fraction of Calcifying Marine Organisms: An *In Vitro* Study of Biomineralization

M. Azizur Rahman[1,*] and Ryuichi Shinjo[2]
[1]*Department of Earth and Environmental Sciences, Palaeontology & Geobiology*
Ludwig-Maximilians-University of Munich
[2]*Department of Physics and Earth Sciences*
University of the Ryukyus, Okinawa
[1]*Germany*
[2]*Japan*

1. Introduction

Only little is known about the early stages of $CaCO_3$ crystallization (Pouget et al. 2009; Gebauer, Volkel, and Colfen 2008), though this mineral has been studied for more than a century now. Identified precursor phases are amorphous calcium carbonate (ACC) in bio- (Addadi, Raz, and Weiner 2003; Weiner et al. 2003) and biomimetic mineralization. Crystal nucleation and biomineralization processes in organisms occur through a sophisticated regulation of internal chemistry that departs significantly from the "constant ionic medium" of seawater (Falini et al. 1996; Addadi et al. 2006; Rahman and Oomori 2009). Magnesium ions are mainly responsible for controlling the kinetics and thermodynamics of calcium carbonate precipitation, especially inhibition of the calcite formation (Davis, Dove, and De Yoreo 2000). The precipitation of calcite at ambient temperature is both thermodynamically and kinetically favored in solutions containing low amounts of magnesium ions. Very recently, it is reported that although Mg^{2+} is influential in producing aragonite in the crystallization process, acidic macromolecules produced calcite crystals in soft corals even in the presence of high Mg^{2+} ions (Rahman, Oomori, and Worheide 2011; Rahman and Oomori 2009).

In the crystallization processes, nucleation is considered to take place in a solution of ions or molecules exceeding a critical supersaturation, leading to the nucleation of the new phase (Addadi, Weiner, and Geva 2001). The growth of nucleated particles and crystals is then considered to take place via addition of single ions or molecules. The role of additives, which modify crystal growth (Rahman and Oomori 2008; Rahman et al. 2011), is restricted in such a view: they can either bind ions or interact with the crystal (Addadi, Berman, Moradianoldak, et al. 1989; Addadi, Weiner, and Geva 2001). It is reported that the crystals

* Corresponding author

grown were completely inhibited and no crystals were observed when a high amount of protein (containing 105 mg/6 ml) was added in the reaction vessel (Rahman and Oomori 2008). The role of additives is, however, still hard to attribute, and empirical control of morphology is the rule, not the exception. This is caused by the multiple roles of additives in such processes, which in addition depends on concentrations and other experimental conditions (Addadi, Berman, Oldak, et al. 1989; Addadi, Weiner, and Geva 2001). A possible quantification, at least a classification of all the different interactions, is eagerly needed, but simple tools to do so are not currently fully known, except few reports (Meldrum and Colfen 2008; Gebauer, Volkel, and Colfen 2008; Rahman, Oomori, and Worheide 2011).

This review aims at opening the pathway to such systematization, here exemplified for calcium carbonate as a model system of complex crystallization. The choice is based on relevance: calcium carbonate is not only of great industrial importance, the major source of water hardness, and the most abundant biominerals, but also one of the most frequently studied minerals, with great scientific relevance in biomineralization and geosciences. Scale formation is also a substantial issue in daily life, industry, and technology, rendering the addition of scale inhibitors to laundry detergents, household cleaners, and also in many industrial applications unavoidable.

2. Mollusks

Biominerals of marine organisms, especially mollusk shells, generally contain unusually acidic proteins (Takeuchi et al. 2008). These proteins are believed to function in crystal nucleation and inhibition. Takeuchi et al (2008) identified an unusually acidic protein Aspein from the pearl oyster Pinctada fucata. They showed that Aspein can control the $CaCO_3$ polymorph (calcite/aragonite) in vitro **(Fig. 1)**. Their results suggest that Aspein is involved in the specific calcite formation in the prismatic layer and the experiments using truncated Aspein demonstrated that the aspartic acid rich domain is crucial for the calcite precipitation. Mollusks, like many other mineralizing organisms, including the vertebrates, first isolate their environment of mineral formation from the outside world (Simkiss 1989). Mollusks use a highly cross-linked protein layer (periostracum) and the epithelial cells of the mantle, the organ directly responsible for shell formation. They then elaborate a matrix within this space comprising various macromolecules (Weiner and Hood 1975; Weiner 1979). This matrix is the framework in which mineral forms (Miyamoto et al. 1996). The major components of the matrix are the polysaccharide beta-chitin, a relatively hydrophobic silk protein, and a complex assemblage of hydrophilic proteins, many of which are unusually rich in aspartic acid (Amos and Evans 2009; Tsukamoto, Sarashina, and Endo 2004; Lowenstam 1989; Samata et al. 1999; Gotliv et al. 2005). The final stage of the process is the formation of the mineral itself within the matrix. Some of the acidic proteins are also occluded within the mineral phase as it forms (Suzuki et al. 2009). The mineral in mature mollusk shells is most often aragonite, sometimes calcite, and in certain taxa, the same shell may have layers of calcite and layers of aragonite (Simkiss 1989; Lowenstam 1989; Suzuki et al. 2009).

Each mineralized tissue contains tens of different macromolecules, many of which are not unique to the mineralized tissue, but can be found in other tissues as well (Arias 2007; Veis 2003; Wilt 2007). There is however one group of glycoproteins that is unique to many

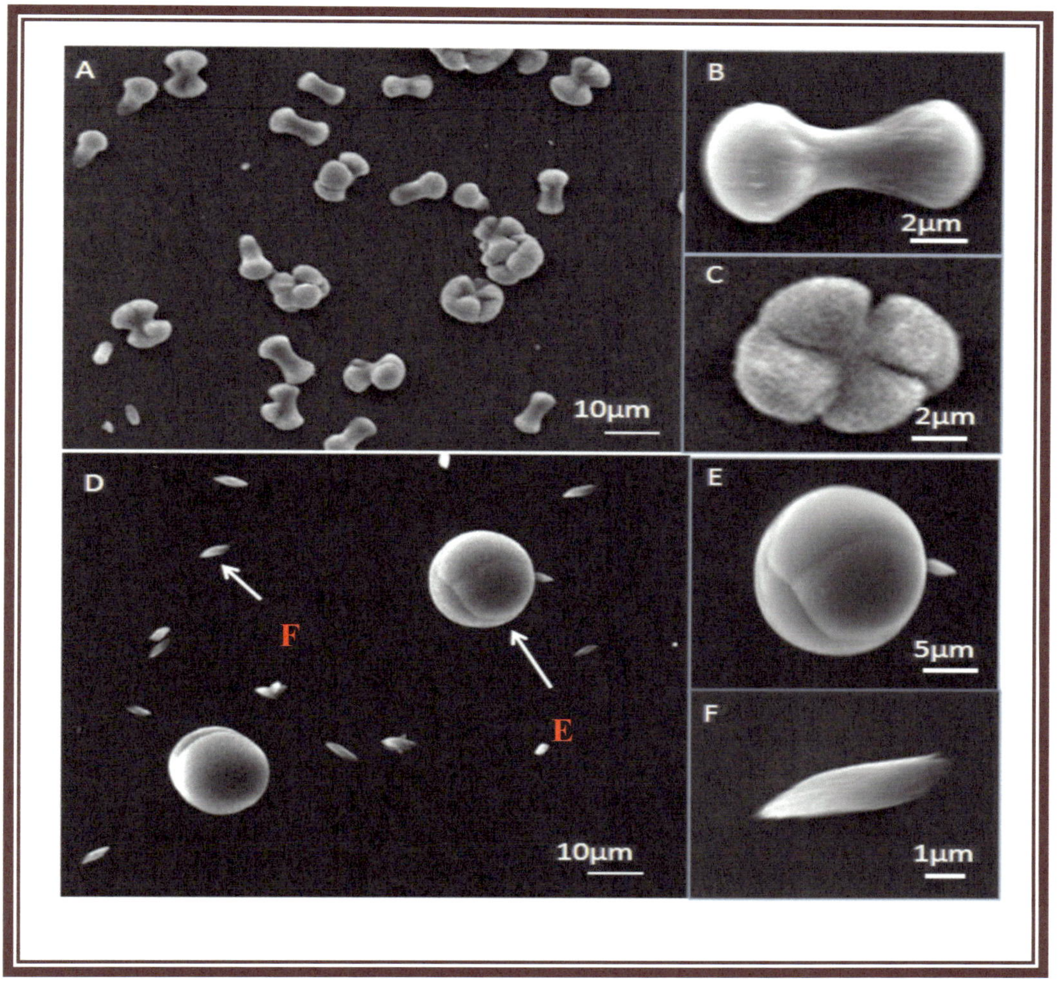

Fig. 1. SEM images of $CaCO_3$ crystals. (A–C) Crystals grown in the presence of Aspein at 10 μg/ml. Dumbbell-like crystals (A, B) are formed. (C) Fused dumbbells are also observed. Under this condition, all the crystals formed are calcite. (D–F) Crystals grown in the presence of Aspein at 2 μg/ml. A spherical crystal (aragonite) is indicated by an arrow in (D) and enlarged in (E). A polyhedral crystal (calcite) is indicated by an arrowhead in (D) and enlarged in (F). From Takeuchi et al. 2008 (with permission)

mineralized tissues. These are proteins that are rich in acidic amino acids, usually aspartic acid (Marin 2007; Weiner 1979). Only a few of these proteins have been sequenced (Weiner 1979; Suzuki et al. 2009), as there are many technical problems in purifying and characterizing such highly charged molecules. Recently, Suzuki, et al. (2009) identified an acidic matrix protein (Pif) in the pearl oyster *Pinctada fucata* that specifically binds to aragonite crystals. The results from immunolocalization, a knockdown experiment that used RNA interference and in vitro calcium carbonate crystallization studies strongly indicate that Pif regulates nacre formation. Others from mollusks have complex domain structures (Sarashina and Endo 1998), including in one case long stretches of polyaspartic acid (Gotliv et al. 2005). These proteins are thought to be the active components of the mineralization process (Marin 2007). They are relatively well investigated in mollusk shell formation. Some are thought to be involved in the formation of the disordered precursor phase, others in the crystal nucleation and growth processes, and others are located within the crystal itself where at least in the case of calcite, they change the materials properties of the crystal (Addadi et al. 2006; Berman et al. 1993; Berman et al. 1990; Nudelman et al. 2007). Such structural proteins do not readily crystallize and the crystal structures of these mineralizing proteins are not yet known. In fact, it seems highly unlikely that they will form crystals, and therefore 3-dimensional structural information may need to be obtained by high resolution cryo-tomography in vitrified ice and/or by solution NMR. Many mineralized tissues fulfill mechanical functions because the presence of mineral causes the tissue to be relatively stiff. Thus understanding the relations between and function often refers to mechanical functions (Addadi and Geva 2003; Addadi and Weiner 1997; Addadi et al. 1995), this is not easy.

Recently, Marie and coworkers (Marie et al. 2008) extracted matrix proteins from a freshwater mussel (*Unio pictorum*) and they found that the nacre-soluble matrix exhibits a carbonic anhydrase activity, an important function in calcification processes. In this study, the shell acid-soluble matrices of prismatic and nacreous layers were prepared. Among the proteins extracted from the shell of *U. pictorum*, P95 was a major component, specific to the nacre acidic soluble matrix (ASM). It was absent in the prismatic layer, unlike the other discrete components. This suggests that P95 might play an important function in controlling, inter alia, the building of nacre during shell formation. P95 is a glycoprotein, the acidity of which is entirely conveyed by its glycosyl moieties, consisting of acidic and sulfated polysaccharides. In its amino acid composition, P95 presents the "signature" of an acidic protein, because of its high Asx and Glx residue content (Weiner 1979). Furthermore, its glycosyl moiety, consisting of sulfated polysaccharides, is involved in calcium binding. To estimate the effects of P95 on the morphology of calcium carbonate crystals, they examined crystals obtained in the presence of P95 by SEM and compared the results with those of a positive control experiment performed with the whole nacre ASM (Marie et al. 2008). Zhang et al. (Zhang, Xie, et al. 2006; Zhang, Li, et al. 2006) isolated a novel matrix protein, designated as p10 from the nacreous layer of pearl oyster (Pinctada fucata) by reverse-phase high-performance liquid chromatography. In vitro crystallization experiments showed that p10 could accelerate the nucleation of calcium carbonate crystals and induce aragonite formation, suggesting that it might play a key role in nacre biomineralization.

3. Vertebrates

One of the best characterized is also the one first identified (Veis 1967), namely phosphophorin extracted from vertebrate teeth (Veis 2003). Often the structures involved, especially those of the vertebrates, are not only hierarchical but also graded—they change in a systematic manner from one location to another (Tesch et al. 2001). One approach for gaining insight into structure function relations in such tissues is to take advantage of the array of surface probe instruments that can provide information on both materials properties and structure at the nanometer level. This indeed has proved to be a powerful approach (Kinney et al. 1999; Moradian-Oldak et al. 2000). A problem with these methods is that it is difficult to integrate the localized information into understanding how a whole organ such as a bone or tooth functions. An alternative approach is to monitor how whole organs deform under load, by mapping the displacements at the nanometer level, and in this way relate them to the structure. This can be done using various optical metrology methods (Shahar 2007). One particularly promising method is electronic speckle pattern interferometry (ESPI), that provides nano-scale deformation information on irregular surfaces even when the object is under water, which is essential for the study of biological tissues (Zaslansky 2006). It is still however a real challenge to integrate structural information at the millimeter, micrometer and nanometer scales and relate this to mechanical properties that are of course the product of all these structures 'working" synergistically. A significant achievement in this regard, is the study by Gupta et al (Gupta et al. 2006) on bone structure–function relations.

An unexpected discovery in the vertebrate biomineralization field was that many tissues of mice in which a minor bone protein, called Matrix Gla Protein (MGP) was removed, spontaneously mineralized (Luo et al. 1997). (Gla or c-carboxyglutamic acid is a most unusual amino acid that resembles glutamic acid, except that it has two carboxylate groups). Clearly one function of MGP is to prevent such catastrophic mineralization. It was also shown that the common serum protein, fetuin-A, has a similar function (Heiss 2002). The calcium phosphate mineral in bones and teeth, carbonated hydroxyapatite, is a relatively insoluble mineral, and there is sufficient calcium and phosphate in vertebrate tissues for them to be saturated with respect to bone mineral. Thus, in the absence of crystal inhibitors, tissues spontaneously mineralize. This led to the interesting proposal that removal of inhibitors is the basic requirement for bone to mineralize (Murshed 2007).

4. Corals

The mechanism by which biomineralization occurs in corals is poorly known. It is reported that corals are composed of calcium carbonate in an organic matrix (Watanabe et al. 2003; Rahman, Oomori, and Worheide 2011; Rahman and Oomori 2008; Rahman et al. 2006; Rahman and Isa 2005; Rahman, Isa, and Uehara 2005; Rahman and Oomori 2008; Fukuda et al. 2003). The organic matrix is formed prior to mineralization, and it has been suggested that some components of the matrix protein may serve as a template for mineral deposition (D'Souza 1999; Weiner and Hood 1975). Recent reports have focused on the characterization of proteins in the soluble matrix of soft coral sclerites (Rahman, Oomori, and Worheide 2011; Rahman and Oomori 2009, 2009, 2008; Rahman, Isa, and Uehara 2005; Rahman et al. 2006; Rahman 2008; Rahman et al. 2011; Rahman and Isa 2005) and stony corals (Watanabe et al. 2003). Also on the control of the morphology and the chemical composition of calcitic bio-crystals in some precious corals have been reported (Dauphin 2006). Compared to the information available on stony corals, molluscans, calcareous algae, and other matrices (Miyamoto et al. 1996; Linde, Lussi, and Crenshaw 1989; Marin, de Groot, and Westbroek 2003; Watanabe et al. 2003; Falini et al. 1996), very little is known regarding matrix components of corals. There are two important features of biomineralization. First, a relatively inert structural frame is built from insoluble macromolecules (hydrophobic proteins, chitin). Second, acidic proteins (rich in aspartic acid, and often in association with sulfated polysaccharides) are assembled on the framework (Mann 1993). It remains important to better understand the role of these matrix fractions in calcification. This can be accomplished by several approaches such as determining how they influence the morphology and composition of the mineral formed.

From the last few years (Rahman et al, 2006, 2009, Watanabe et al 2003), acidic proteins were purified from the organic matrices of corals. In the present review, the crystallization of corals in the presence of these proteins was discussed and compared with other macromolecules. Crystallization plays a key role in the bio-calcification process and ultimately in the growth of coral skeletons. One widely used approach for studying the functions of these acidic proteins is to examine their effect on crystal growth, in vitro. Combinations of matrix components have been used to detect a collaborative effect (Termine et al. 1981), and the ability of demineralized matrix to induce crystal nucleation has been examined (Rahman, Oomori, and Worheide 2011). The objective was to understand the principles that govern these interactions and to gain insight into the mechanisms by which these matrix constituents regulate crystal growth in vivo. The major polymorphisms involved in $CaCO_3$ crystallization of marine organisms were identified and subsequently, the functions of specific organic matrix proteins in the bio-calcification process were determined.

The similar results were found in the soft corals sclerites **(Fig. 2).** To investigate the influence of matrix proteins on calcium carbonate crystals *in vitro*, the morphology of crystals grown with or without any protein was observed under SEM and XRD by Rahman et al. **(Fig. 3)**. Two crystallization solutions were prepared, one that induces calcite (calcitic crystallization solution) and one that induces aragonite (aragonitic crystallization solution). The calcitic solution was a supersaturated solution of $Ca(HCO_3)_2$ prepared by purging a stirred aqueous suspension of $CaCO_3$ with carbon dioxide.

The crystals grown in the absence of protein exhibited the characteristic rhombohedral morphology of calcite **(Fig. 2A).** However, crystals grown in the presence of matrix protein showed an interesting phenomenon. In comparison to the control measurement, crystal growth density was lower when a high amount of matrix (containing 45 μg/6ml protein) was added into the reaction vessel **(Fig. 2B).** The morphology of crystals was also affected in the presence of 45 μg of protein (see enlarged image of the boxed part in **Fig. 2C**). Subsequently, when 1/10 of the amount of matrix (containing 4.5 μg/6ml protein) was added to the solutions at the same time, crystal growth increased and a few crystals were affected **(Fig. 2D).** No crystals were observed under SEM when an even higher amount of protein (containing 105 μg/6ml protein) was added in the reaction vessel **(Fig. 2E).** These results demonstrate the regulation of crystal growth by protein in a $CaCO_3$ system in alcyonarian corals, specifically the inhibition of crystal growth. This study concurred with the similar works recently conducted by Takeuchi and coworkers (Takeuchi et al. 2008; Suzuki et al. 2009).

Further, precipitation of $CaCO_3$ was simulated *in vitro* in the presence of both soluble and insoluble organic matrix proteins of sclerites at the ratio of 29:60 mol% **(Fig. 3).** In every experiment, influence of proteins with the precipitated crystals was examined by XRD. **Table 1** summarizes the percentage of aspartic acid and other amino acids in the soluble and insoluble matrix proteins used for crystallization. **Figure 3** shows the $CaCO_3$ crystal growth and morphology in the absence or presence of matrix proteins. The influence of Mg^{2+} on $CaCO_3$ polymorphism was also studied. Without Mg^{2+}, typical rhombohedral calcite crystals were generated **(Fig. 3A, B and L).** In the presence of Mg^{2+} (50 mM), large needle-like crystals were preferentially formed **(Fig. 3C, D and M)**. **Fig. 3 (E–K)** shows the SEM images

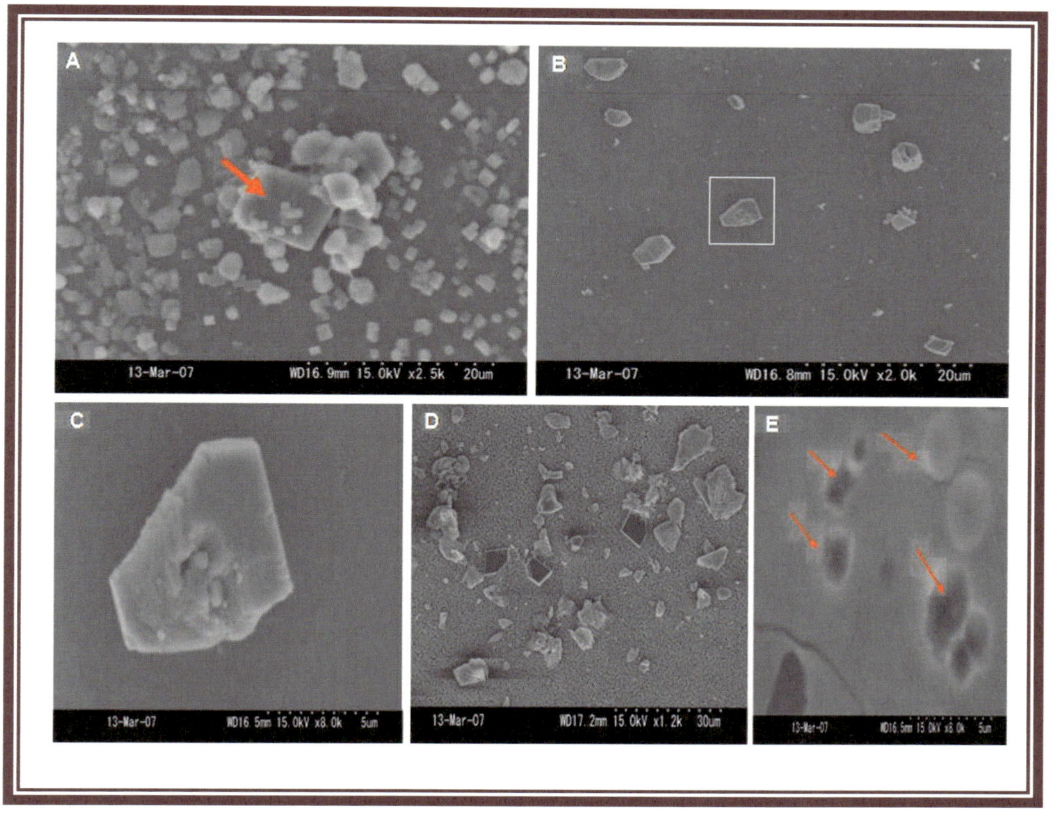

Fig. 2. SEM observation of crystals in the presence or absence of matrix proteins isolated from the calcitic sclerites. (A) Calcite rhombohedrons grown in the absence of matrix proteins, the arrow indicates the {104} face of the crystals. (B) Calcite crystals grown in the presence of matrix proteins (protein concentration 45 mg/6 ml). (C) Enlarged image of the boxed part in (B). (D) Calcite crystals grown in the presence of matrix proteins (protein concentration 4.5 mg/6 ml). (E) Result of a crystallization experiment that was carried out by the same procedure as in (B) and (D) but in the presence of high amount matrix proteins (containing 105 mg/6 ml). There is no mineral deposition (also at higher magnification), i.e., only the glass spot or very tiny crystals are seen (arrows), indicating that the crystallization is completely inhibited by the high content of matrix proteins. Scale bars: A= 20 µm; B=20 µm; C= 5 µm; D=30 µm; E=5µm. From Rahman and Oomori 2008.

of $CaCO_3$ crystals grown in the presence of matrix protein, at a lower and a higher concentration, 0.5 and 1.4 µg/mL, respectively. At a concentration of 0.5 µg/mL, a number of spherical crystals (aragonite) remained **(Fig. 3E indicated by an arrowhead, F),** and some polyhedral and round calcite crystals were exclusively induced (**Fig. 3E indicated by arrows, G, H)**. The XRD measurements **(Fig.3N)** demonstrated that the both aragonite and calcite crystals are available in this experiment. The higher concentration of matrix proteins (1.4 µg/mL) showed a high intensity of rhombohedral calcite crystals **(Fig. 3I, J, K)** without any aragonite formation. Although the growth of crystals was inhibited at high concentrations of proteins, all remaining aragonites formed by Mg^{2+} (50 mM) were transformed into calcites **(Fig. 3I and enlarged view indicated by arrows in J, K).** The XRD measurements proved that all crystals formed under these conditions were calcites **(Fig. 3O)**. These observations strongly suggest that the acidic matrix proteins are the key components in forming calcite crystals in biocalcification.

From our observation, the density of nucleation sites was lower when the crystals were grown with a mixture of soluble and insoluble matrix proteins in which the aspartic acid content of the insoluble matrix was 60mol% **(Fig. 3I).** Also, soft corals have special characters because the organic matrices themselves are highly aspartic acid-rich proteins (Rahman and Oomori 2009). In addition, previous studies on molluscan shells indicate that acidic amino acid residues may actually inhibit crystal nucleation (Wilbur 1982). The present review reveals that both matrix fractions of some marine organisms were enriched in aspartic acid proteins (Weiner 1979; Takeuchi et al. 2008; Rahman and Oomori 2009); the especially high aspartic acid content of insoluble organic matrix proteins may regulate the crystal growth and morphology (Suzuki et al. 2009; Rahman et al. 2006; Rahman 2008), or could play a key role in crystal nucleation induction (Addadi et al. 2006; Addadi, Raz, and Weiner 2003; Addadi et al. 1995; Addadi, Berman, Oldak, et al. 1989).

	Soluble fraction		Insoluble fraction	
Amino acid	Molecular weight of residues	mol%	Molecular weight of residues	mol%
Cys	151.14	0.41	151.14	0.25
Asx (Asp+Asn)	115.09	29.35	115.09	60.92
Thr	101.10	5.18	101.10	1.57
Ser	87.08	4.61	87.08	1.52
Glx (Glu+Gln)	129.11	9.63	129.11	5.30
Gly	57.05	10.37	57.05	9.40
Ala	71.08	11.37	71.08	12.87
Val	99.13	5.06	99.13	1.26
1/2-Cys	103.14	1.85	103.14	0.55
Met	131.20	0.82	131.20	0.17
Ile	113.16	2.99	113.16	0.74
Leu	113.16	3.56	113.16	1.05
Tyr	163.17	0.80	163.17	0.32
Phe	147.17	1.93	147.17	0.51
Lys	128.17	2.66	128.17	0.79
His	137.14	0.78	137.14	0.29
Arg	156.19	1.89	156.19	0.92
Trp	186.20	0.00	186.20	0.00
Pro	97.11	6.73	97.11	1.60
Protein Total		100.00		100.00

Table 1. Amino acid composition of the protein in the total soluble and insoluble fractions from the calcitic sclerites of *S. polydactyla*. From Rahman and Oomori 2009

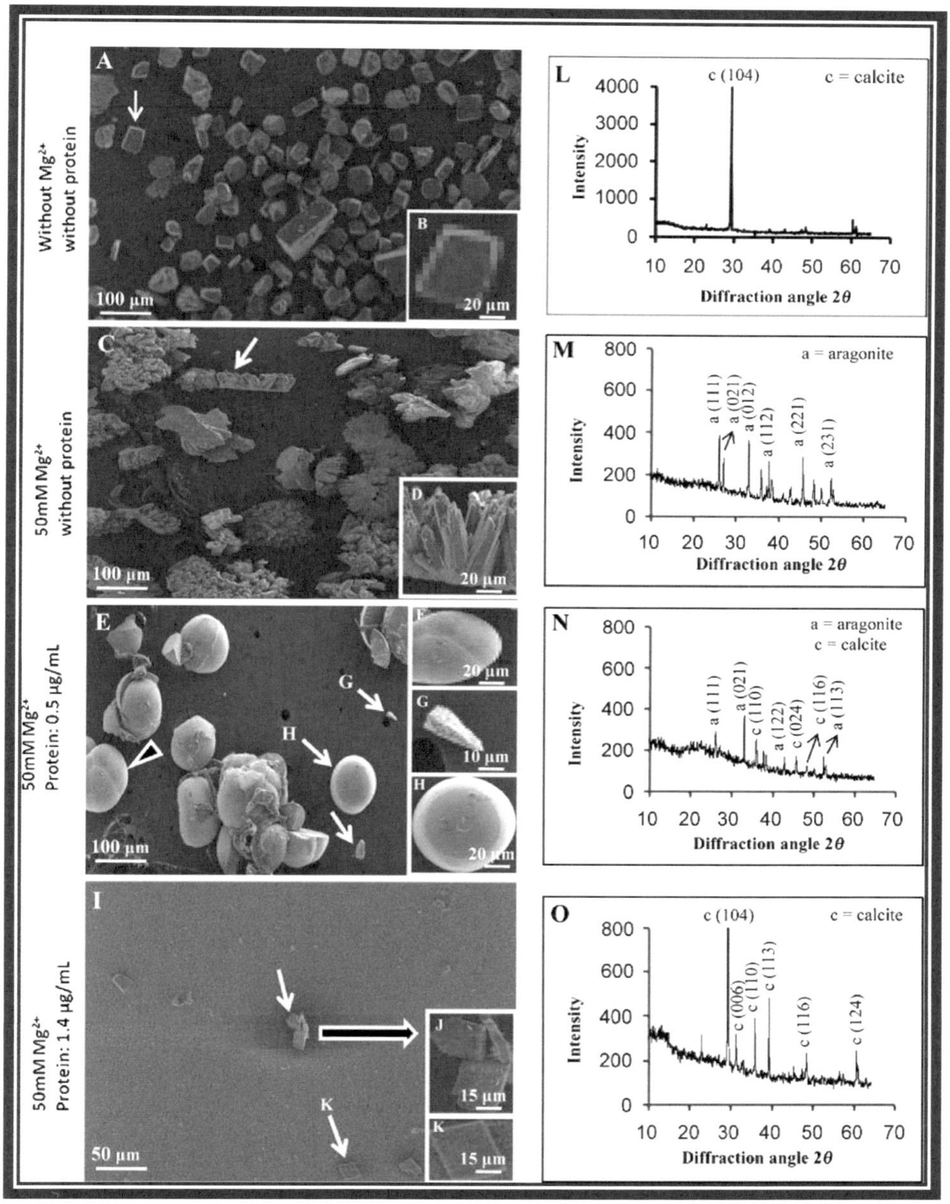

Fig. 3. SEM images of CaCO3 crystals growth *in vitro* and their XRD diffractions. A mixture of soluble and insoluble organic matrix proteins (insoluble = 60% aspartic acid; soluble = 29% aspartic acid) isolated from the calcitic sclerites was used in crystallization experiments. (A, B) Crystals grown without proteins in the absence of Mg2+ shows rhombohedral calcite crystals. (C, D) Crystals grown without proteins in the presence of Mg2+ shows needle-like crystals (aragonite). (E–H) Crystals grown in the presence of proteins (0.5 μg/mL) with Mg2+ (50 mM). A number of spherical crystals (aragonite) remained (indicated by an arrowhead in E and enlarged in F), and some polyhedral and round calcite crystals were exclusively induced

(indicated by arrows in E and enlarged in G, and H). The polymorphs of CaCO3 were identified both by XRD and Raman microprobe analysis (see Figure 3 for Raman data). (I–K) Crystals grown in the presence of proteins (1.4 μg/mL) with Mg2+ (50 mM). Under these conditions, all the crystals formed are calcites (arrows in I and enlarged in J and K). (L–O) Verification of crystals formed in these experiments by XRD. From Rahman and Oomori 2009

5. Concluding remarks

The overall aim of this review is to better understand the function of the acidic proteins in the matrices of marine organisms during mineral formation by using an *in vitro* analysis approach. Since it is difficult to understand biological systems based on *in vivo* analyses of control and regulation processes, the actions of additives such as acidic proteins, which are mostly available in calcifying marine organisms, have been discussed here based on *in vitro* analyses. *In vitro* experiments reported in the literature reveal that acidic proteins are primarily responsible for the control of $CaCO_3$ polymorphisms in mollusk shells and corals. In this paper, we used data from *in vitro* experiments with mollusk shells and corals, since these marine organisms contain high concentrations of acidic proteinaceous fractions. We compared the data from vertebrate studies to evaluate the differences between calcifying marine organisms (mollusk shells and corals). However, because of the different characterization of proteinaceous fractions in vertebrates, polymorphism control and regulation processes are completely different among invertebrate calcifying marine organisms. We conclude that there is potential for the control of $CaCO_3$ precipitation in calcifying marine organisms, which contain a high content of proteinaceous fractions. Our review suggests that the acidic matrix proteins in calcifying marine organisms are a specialized nucleating sheet that governs the nucleation of highly-oriented calcite or aragonite crystals and, thus, lead to a finer crystallization of $CaCO_3$.

6. Acknowledgments

This work was supported by the Alexander von Humboldt Foundation, Germany.

7. References

Addadi, L., J. Aizenberg, S. Albeck, G. Falini, and S. Weiner. 1995. Structural control over the formation of calcium carbonate mineral phases in biomineralization. *Supramolecular Stereochemistry* 473:127-139

Addadi, L., A. Berman, J. Moradianoldak, and S. Weiner. 1989. Protein-Crystal Interactions in Biomineralization. *Abstracts of Papers of the American Chemical Society* 197:115-IAEC.

Addadi, L., A. Berman, J. M. Oldak, and S. Weiner. 1989. Structural and stereochemical relations between acidic macromolecules of organic matrices and crystals. *Connective Tissue Research* 21 (1-4):127-34; discussion 135.

Addadi, L., and M. Geva. 2003. Molecular recognition at the interface between crystals and biology: generation, manifestation and detection of chirality at crystal surfaces. *Crystengcomm*:140-146.

Addadi, L., D. Joester, F. Nudelman, and S. Weiner. 2006. Mollusk shell formation: A source of new concepts for understanding biomineralization processes. *Chemistry-a European Journal* 12 (4):981-987.

Addadi, L., S. Raz, and S. Weiner. 2003. Taking advantage of disorder: Amorphous calcium carbonate and its roles in biomineralization. *Advanced Materials* 15 (12):959-970.

Addadi, L., and S. Weiner. 1997. Biomineralization - A pavement of pearl. *Nature* 389 (6654):912-&.

Addadi, L., S. Weiner, and M. Geva. 2001. On how proteins interact with crystals and their effect on crystal formation. *Z Kardiol* 90 Suppl 3:92-8.

Amos, F. F., and J. S. Evans. 2009. AP7, a partially disordered pseudo C-RING protein, is capable of forming stabilized aragonite in vitro. *Biochemistry* 48 (6):1332-9.

Arias, J.L., Ferna´ndez, M.S. 2007. Biomineralization: From Paleontology to Materials Science. *Editorial Universitaria, Santiago.*

Berman, A., L. Addadi, A. Kvick, L. Leiserowitz, M. Nelson, and S. Weiner. 1990. Intercalation of sea urchin proteins in calcite: study of a crystalline composite material. *Science* 250 (4981):664-7.

Berman, A., J. Hanson, L. Leiserowitz, T. F. Koetzle, S. Weiner, and L. Addadi. 1993. Biological control of crystal texture: a widespread strategy for adapting crystal properties to function. *Science* 259 (5096):776-9.

D'Souza, S. M. Alexander, C. Carr, S. W Waller, A. M.. Whitcombe, M. J Vulfson, E. N. 1999. Directed nucleation of calcite at a crystal-imprinted polymer surface. *Nature* 319:312-316.

Dauphin, Y. 2006. Mineralizing matrices in the skeletal axes of two Corallium species (Alcyonacea). *Comp Biochem Physiol A Mol Integr Physiol* 145 (1):54-64.

Davis, K. J., P. M. Dove, and J. J. De Yoreo. 2000. The role of Mg2+ as an impurity in calcite growth. *Science* 290 (5494):1134-1137.

Falini, G., S. Albeck, S. Weiner, and L. Addadi. 1996. Control of aragonite or calcite polymorphism by mollusk shell macromolecules. *Science* 271 (5245):67-69.

Fukuda, I., S. Ooki, T. Fujita, E. Murayama, H. Nagasawa, Y. Isa, and T. Watanabe. 2003. Molecular cloning of a cDNA encoding a soluble protein in the coral exoskeleton. *Biochem Biophys Res Commun* 304 (1):11-7.

Gebauer, D., A. Volkel, and H. Colfen. 2008. Stable prenucleation calcium carbonate clusters. *Science* 322 (5909):1819-22.

Gotliv, B. A., N. Kessler, J. L. Sumerel, D. E. Morse, N. Tuross, L. Addadi, and S. Weiner. 2005. Asprich: A novel aspartic acid-rich protein family from the prismatic shell matrix of the bivalve Atrina rigida. *Chembiochem* 6 (2):304-14.

Gupta, H. S., J. Seto, W. Wagermaier, P. Zaslansky, P. Boesecke, and P. Fratzl. 2006. Cooperative deformation of mineral and collagen in bone at the nanoscale. *Proc Natl Acad Sci U S A* 103 (47):17741-6.

Heiss, A., DuChesne,A.,Denecke, B.,Gro¨tzinger, J.,Yamamoto,K.,Renne´, T., Jahnen-Duchent,W. 2002. Structural basis of calcification inhibition by a2 HS glycoprotein/fetuin-A. *J. Biol. Chem.* 278:13333-13341.

Kinney, J. H., M. Balooch, G. W. Marshall, and S. J. Marshall. 1999. A micromechanics model of the elastic properties of human dentine. *Arch Oral Biol* 44 (10):813-22.

Linde, A., A. Lussi, and M. A. Crenshaw. 1989. Mineral induction by immobilized polyanionic proteins. *Calcif Tissue Int* 44 (4):286-95.

Lowenstam, H. A, Weiner, S. 1989. On Biomineralization. *Oxford University Press, New York.*

Luo, G., P. Ducy, M. D. McKee, G. J. Pinero, E. Loyer, R. R. Behringer, and G. Karsenty. 1997. Spontaneous calcification of arteries and cartilage in mice lacking matrix GLA protein. *Nature* 386 (6620):78-81.

Mann, S. 1993. Molecular Tectonics in Biomineralization and Biomimetic Materials Chemistry. *Nature* 365 (6446):499-505.

Marie, B., G. Luquet, L. Bedouet, C. Milet, N. Guichard, D. Medakovic, and F. Marin. 2008. Nacre calcification in the freshwater mussel Unio pictorum: carbonic anhydrase activity and purification of a 95 kDa calcium-binding glycoprotein. *Chembiochem* 9 (15):2515-23.

Marin, F., K. de Groot, and P. Westbroek. 2003. Screening molluscan cDNA expression libraries with anti-shell matrix antibodies. *Protein Expr Purif* 30 (2):246-52.

Marin, F., Luquet, G. 2007. Unusually acidic proteins in biomineralization. *Handbook of Biomineralization. Biological Aspects and Structure Formation. Wiley VCH, Weinheim*:273-290.

Meldrum, F. C., and H. Colfen. 2008. Controlling mineral morphologies and structures in biological and synthetic systems. *Chem Rev* 108 (11):4332-432.

Miyamoto, H., T. Miyashita, M. Okushima, S. Nakano, T. Morita, and A. Matsushiro. 1996. A carbonic anhydrase from the nacreous layer in oyster pearls. *Proc Natl Acad Sci U S A* 93 (18):9657-60.

Moradian-Oldak, J., M. L. Paine, Y. P. Lei, A. G. Fincham, and M. L. Snead. 2000. Self-assembly properties of recombinant engineered amelogenin proteins analyzed by dynamic light scattering and atomic force microscopy. *Journal of Structural Biology* 131 (1):27-37.

Murshed, M., Hamey, D., Milla´n, J.L., McKee, M.D., Karsenty, G. 2007. Unique expression in osteoblasts of broadly expressed genes accounts for the spatial restriction of ECM mineralization to bone. *Genes Dev.* 19:1093-1104.

Nudelman, F., H. H. Chen, H. A. Goldberg, S. Weiner, and L. Addadi. 2007. Spiers Memorial Lecture. Lessons from biomineralization: comparing the growth strategies of mollusc shell prismatic and nacreous layers in Atrina rigida. *Faraday Discuss* 136:9-25; discussion 107-23.

Pouget, E. M., P. H. Bomans, J. A. Goos, P. M. Frederik, G. de With, and N. A. Sommerdijk. 2009. The initial stages of template-controlled CaCO3 formation revealed by cryo-TEM. *Science* 323 (5920):1455-8.

Rahman, M. A., H. Fujimura, R. Shinjo, and T. Oomori. 2011. Extracellular matrix protein in calcified endoskeleton: a potential additive for crystal growth and design. *Journal of Crystal Growth* 324 (1):177-183.

Rahman, M. A., and Y. Isa. 2005. Characterization of proteins from the matrix of spicules from the alcyonarian, Lobophytum crassum. *Journal of Experimental Marine Biology and Ecology* 321 (1):71-82.

Rahman, M. A., Y. Isa, A. Takemura, and T. Uehara. 2006. Analysis of proteinaceous components of the organic matrix of endoskeletal sclerites from the alcyonarian Lobophytum crassum. *Calcified Tissue International* 78 (3):178-185.

Rahman, M. A., Y. Isa, and T. Uehara. 2005. Proteins of calcified endoskeleton: II Partial amino acid sequences of endoskeletal proteins and the characterization of proteinaceous organic matrix of spicules from the alcyonarian, Synularia polydactyla. *Proteomics* 5 (4):885-893.

Rahman, M. A., and T. Oomori. 2008. Identification and Function of New Proteins in Calcified endoskeleton: a New Insight in the Calcification Mechanism of Soft Corals. *Oceans 2008, Vols 1-4*:2139-2145

Rahman, M. A., and T. Oomori. 2008. Structure, crystallization and mineral composition of sclerites in the alcyonarian coral. *Journal of Crystal Growth* 310 (15):3528-3534.

Rahman, M. A., and T. Oomori. 2009. Analysis of Protein-induced Calcium Carbonate Crystals in Soft Coral by Near-Field IR Microspectroscopy. *Analytical Sciences* 25 (2):153-155.

Rahman, M. A., and T. Oomori. 2009. In Vitro Regulation of CaCO3 Crystal Growth by the Highly Acidic Proteins of Calcitic Sclerites in Soft Coral, Sinularia Polydactyla. *Connective Tissue Research* 50 (5):285-293.

Rahman, M. A., T. Oomori, and G. Worheide. 2011. Calcite formation in soft coral sclerites is determined by a single reactive extracellular protein. *J Biol Chem* 286 (36):31638-49.

Rahman, M. A, T. Oomori, 2008. Aspartic Acid-rich Proteins in Insoluble Organic Matrix Play a Key Role in the Growth of Calcitic Sclerites in Alcyonarian Coral. *Chin J Biotech* 24 (12):2127−2128.

Samata, T., N. Hayashi, M. Kono, K. Hasegawa, C. Horita, and S. Akera. 1999. A new matrix protein family related to the nacreous layer formation of Pinctada fucata. *Febs Letters* 462 (1-2):225-229.

Sarashina, I., and K. Endo. 1998. Primary structure of a soluble matrix protein of scallop shell: Implications for calcium carbonate biomineralization. *American Mineralogist* 83 (11-12):1510-1515.

Shahar, R., Weiner, S. 2007. Insights into whole bone and tooth function using optical metrology. *J. Mat. Sci.* 42:8919-8933.

Simkiss, K. Wilbur, K. . 1989. Biomineralization. *Cell Biology and Mineral Deposition, Academic Press, San Diego.*

Suzuki, M., K. Saruwatari, T. Kogure, Y. Yamamoto, T. Nishimura, T. Kato, and H. Nagasawa. 2009. An acidic matrix protein, Pif, is a key macromolecule for nacre formation. *Science* 325 (5946):1388-90.

Takeuchi, T., I. Sarashina, M. Iijima, and K. Endo. 2008. In vitro regulation of CaCO(3) crystal polymorphism by the highly acidic molluscan shell protein Aspein. *FEBS Lett* 582 (5):591-6.

Termine, J. D., H. K. Kleinman, S. W. Whitson, K. M. Conn, M. L. McGarvey, and G. R. Martin. 1981. Osteonectin, a bone-specific protein linking mineral to collagen. *Cell* 26 (1 Pt 1):99-105.

Tesch, W., N. Eidelman, P. Roschger, F. Goldenberg, K. Klaushofer, and P. Fratzl. 2001. Graded microstructure and mechanical properties of human crown dentin. *Calcif Tissue Int* 69 (3):147-57.

Tsukamoto, D., I. Sarashina, and K. Endo. 2004. Structure and expression of an unusually acidic matrix protein of pearl oyster shells. *Biochem Biophys Res Commun* 320 (4):1175-80.

Veis, A. 2003. Mineralization in organic matrix frameworks. *Biomineralization* 53:250-290.

Veis, A., Perry, A. 1967. The phosphoprotein of the dentin matrix. *Biochemistry* 6:2409–2416.

Watanabe, T., I. Fukuda, K. China, and Y. Isa. 2003. Molecular analyses of protein components of the organic matrix in the exoskeleton of two scleractinian coral species. *Comp Biochem Physiol B Biochem Mol Biol* 136 (4):767-74.

Weiner, S. 1979. Aspartic acid-rich proteins: major components of the soluble organic matrix of mollusk shells. *Calcif Tissue Int* 29 (2):163-7.

Weiner, S., and L. Hood. 1975. Soluble-Protein of Organic Matrix of Mollusk Shells - Potential Template for Shell Formation. *Science* 190 (4218):987-988.

Weiner, S., Y. Levi-Kalisman, S. Raz, and L. Addadi. 2003. Biologically formed amorphous calcium carbonate. *Connective Tissue Research* 44:214-218.

Wilbur, K. M, Bernhardt, A.M. 1982. Mineralization of molluscan shell: effects of free and polyamono acids on crystal growth rate in vitro. *Am. Zool.* 22:952.

Wilt, F., Ettensohn, C.A. 2007. The morphogenesis and biomineralization of the sea urchin larval skeleton. *Handbook of Biomineralization. Biological Aspects and Structure Formation. Wiley VCH, Weinheim*:183-210.

Zaslansky, P., Shahar, R., Friesem, A.A., Weiner, S. 2006. Relations between shape, materials properties and function in biological materials using laser speckle interferometry: in situ tooth deformation. *Adv. Funct. Mat.* 16:1925-1936.

Zhang, C., S. Li, Z. Ma, L. Xie, and R. Zhang. 2006. A novel matrix protein p10 from the nacre of pearl oyster (Pinctada fucata) and its effects on both CaCO3 crystal formation and mineralogenic cells. *Mar Biotechnol (NY)* 8 (6):624-33.

Zhang, C., L. Xie, J. Huang, X. Liu, and R. Zhang. 2006. A novel matrix protein family participating in the prismatic layer framework formation of pearl oyster, Pinctada fucata. *Biochem Biophys Res Commun* 344 (3):735-40.

Part 2

In vivo Mineralization Systems

4

The Chiton Radula: A Unique Model for Biomineralization Studies

Lesley R. Brooker[1] and Jeremy A. Shaw[2]
[1]*University of the Sunshine Coast*
[2]*Centre for Microscopy, Characterisation & Analysis*
University of Western Australia
Australia

1. Introduction

Over the course of evolution, a range of strategies have been developed by different organisms to produce unique materials and structures perfected for their specific function. This biological mastery of materials production has inspired the birth of the new discipline of biomaterials through biomimicry (Birchall, 1989).

Chitons (Mollusca: Polyplacophora) are slow moving, bilaterally symmetrical and dorso-ventrally flattened molluscs that are commonly found on hard substrata in intertidal regions of coastlines around the world (Kaas & Jones, 1998). All species are characterized by a series of eight dorsal, articulating shell plates or valves, which may be embedded, to varying degrees, in a fleshy, muscular girdle (Kaas & Jones, 1998) (Figure 1). Approximately 750 living species are known, and while intertidal regions are home to the majority of chitons, a number of species can be found at depths of up to 8000m where they feed on detrital material (Kaas & Jones, 1998).

Fig. 1. Photograph of the dorsal surface of the chiton *Acanthopleura gaimardi*, showing the eight overlapping aragonite plates surrounded by the fleshy girdle, which, in this species, is covered in small aragonite spines.

Chitons feed by rasping macro- and micro-algae from the rocks on which they live through the use of a radula. The radula has been coined as a conveyor belt of continuously developing teeth, replaced by new teeth as they are worn and lost. The chiton radula is an exquisite example of nature at its best, where matrix mediated biomineralization controls the deposition of a wide range of minerals in architecturally discrete regions resulting in highly specialized feeding implements. This deposition of the biominerals within an organic framework facilitates intricate crystallographic design and structure, and imparts unique properties to the chiton teeth, such as tensile strength, shock absorption and controlled wear and abrasion (Wealthall et al., 2005). The biomineralized teeth of chitons are sophisticated composite structures that have been refined by evolution over millions of years, resulting in highly efficient, self-sharpening, feeding implements ideally suited to their function. As such, the resultant biologically optimized tools possess many of the desirable features we seek in new and innovative biomaterials. Indeed, the unique structure of chiton teeth inspired the design of dredging equipment, providing an example of engineering biomimicry (van der Wal et al., 2000). The chiton radula is a highly appropriate model to use in biomineralization studies because of its potential for the development of new and innovative biomaterials, and biomimicry platforms for industrial and biomedical applications.

In chiton radula teeth, a range of iron oxides, including magnetite, is deposited under ambient temperature and pressure. This is in stark contrast to that which is achievable in current industrial processes, where extreme temperature and pressure conditions, combined with low oxygen levels, have to be maintained for magnetite production. An understanding of the biological parameters that facilitate magnetite formation in the chiton radula is imperative to be able to replicate this phenomenon in industrial and biomaterials applications. In addition to the iron minerals, the chiton mineralizes its radula teeth with $CaPO_4$, in stark contrast to most other invertebrates that mineralize their hard structures with $CaCO_3$. Since this is the same mineral found in bones and teeth, an understanding of the processes governing its deposition has direct relevance to and significance for, bone tissue and dental technologies. Hence, the chiton model offers exciting opportunities for application in both medical and industrial biomaterials contexts.

An understanding of the processes of mineral deposition in chiton teeth is fundamental to our wider understanding of biomineralization processes, and highly relevant to a modern materials technology focus, a rapidly developing area that, through nanotechnology and associated crystal design, is already revolutionizing our everyday world.

1.1 Uniqueness of the chiton radula for biomineralization studies

Unlike most other biomineral systems, the chiton radula presents a complete temporal and spatial story of biomineralization. In mollusc shell or vertebrate bone, for instance, biomineralization is a dynamic process with continuous formation, and even remodeling, of the structure through deposition of alternating layers of organic matrix and calcium mineral. As such, it is difficult to distinguish between the different processes. In contrast, chiton teeth are fabricated in a manner resembling a production line, with each successive tooth row steadily progressing in stages from the unmineralized to the mineralized state (Macey & Brooker, 1996). Hence, present in this one tissue, are all stages of the biomineralization process: the initial development and maturation of the organic scaffold; the cellular delivery of ions to the matrix; the deposition of the precursor iron molecules; the highly organized and sequential deposition of a range of iron minerals; the infilling of the core of the tooth with amorphous calcium phosphate granules; and the conversion to crystalline calcium phosphate. This assembly line of biomineralization in chiton teeth (Figure 2) has facilitated detailed examination of every step of the process, providing a unique insight into many of the fundamental principles governing biomineralization in organisms.

Fig. 2. Light micrograph of the radula of the chiton *Acanthopleura gaimardi* showing the progressive stages of radular tooth development. From the clear unmineralized teeth, comprised of a chitinous organic matrix on the right, to the black, fully mineralized, working teeth on the left.

1.2 Barriers to studying biomineralization in the chiton radula

The structural properties of the chiton radula have presented many challenges to researchers, which have required ingenuity, a multidisciplinary approach and the application of novel techniques and methodologies to overcome. The radula is constructed of a range of minerals varying in hardness with magnetite at the extreme, having a Mohs score of 6, hydroxyapatite at 5, an organic matrix that is relatively soft and surrounding tissue that offers little structural resistance. While a superb example of bioengineering, the chiton radula has presented an almost insurmountable obstacle to the use of conventional histological techniques to examine all components of the radula teeth *in situ*. With the difficulties associated with processing the radula intact, many of the earlier studies either attempted to separate the component parts of the radula and undertake bulk analysis or stripped away the mineral component to examine the organics, or the organic component to analyze the mineral. Subsequent studies that attempted to examine the radula teeth *in situ* were limited to an examination of the pre-mineralization region of the radula and extrapolation of the results to surmise what actually happens in the mineralized teeth, or to designing *in vitro* experiments that may mimic the natural processes. With the advent and adoption of more sophisticated methodologies and techniques, it has finally become possible to examine the fully mineralized teeth of chitons *in situ*.

2. Morphology of the chiton radula

The radula is a feeding organ common to all molluscs, with the exception of the bivalves which are filter feeders. The chiton radula extends back from the mouth to approximately one third of the animal's length and lies within a sac of tissue that directs radula development. The radula sac is divided into three broad functional groups of cells: the odontoblast cells at the posterior end, which are responsible for producing the teeth (Nesson & Lowenstam, 1985; Eernisse & Kerth, 1988;); the inferior epithelium, which carries the developing radula forward towards the mouth; and the superior epithelium, which supplies the molecules and ions required for mineralization of the teeth (Shaw et al., 2009a). The radula teeth sit on a chitinous radular membrane, which is fused at the anterior end to a cuticular structure, the subradular membrane (Fretter & Graham, 1962; Graham, 1973). The immature radula undergoes a complex suite of maturation processes as it progresses anteriorly. While at the mature end of the radula, teeth that are lost through wear and breakage during feeding are continually replaced by newly developed teeth (Runham, 1962; Shaw et al., 2002, 2008a). In the daily process of feeding, the radula teeth of chitons are subject to highly abrasive conditions and meet this challenge by replacing each transverse row of teeth every two to three days (Shaw et al., 2002, 2008a)

The chiton radula is bilaterally symmetrical around a central rachidian tooth, with (usually) 17 teeth to each transverse row (Figure 3). It is polystichous, since there are many different teeth in each row, and serially repeated, since all rows are composed of the same tooth arrangement, with from 25 to 150 rows of teeth, depending on the species (Eernisse & Reynolds, 1994; Macey & Brooker, 1996; Brooker & Macey, 2001; Shaw et al. 2002, 2008a; Brooker, 2004; Brooker et al., 2006).

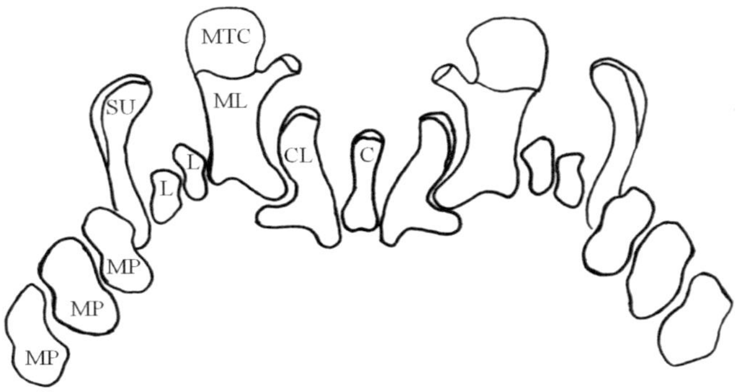

Fig. 3. Diagram of a typical row of teeth in the radula of the genus *Acanthopleura*. **C**, central; **CL**, centrolateral; **L**, lateral plates; **ML**, major lateral; **MP**, marginal plates; **MTC**, mineralized tooth cusps; **SU**, spatulate uncinal.

The largest teeth in each row, the major laterals, are easily distinguished by their glossy black cusps, due to their impregnation with magnetite. The major lateral teeth consist of a base, a shaft (stylus) and a cusp (Kaas & Jones, 1998), with the region between the cusp and shaft referred to as the junction zone (Macy & Brooker, 1996). The shape of the cusp is species specific, varying from a simple broad disc or shovel to being attenuated into a point, and they may possess from two to four denticles (Brooker & Macey, 2001). While various metal ions have been found throughout the major lateral teeth, minerals are restricted to the cusp (Macey & Brooker, 1996).

3. Structure and composition of the chiton radula

3.1 Matrix composition of chiton teeth

The chemical composition of the molluscan radula was the subject of investigation as early as the 17th Century when Leuckart (in Sollas, 1907) determined that the gastropod and cephalopod radulae were composed of the polysaccharide chitin, which was subsequently confirmed for all odontophorous molluscs (Sollas, 1907). Evans et al. (1990) confirmed α-chitin to be the principal component of the tooth matrix in the chiton *Acanthopleura hirtosa*, and further showed that it consisted of 10% (by weight) proteins, which were rich in aspartic and glutamic amino acids (Evans et al., 1991). It is the presence of acidic proteins in the matrix of biomineralized structures that Addadi & Weiner (1985) identified as fundamental to the initiation of crystal formation. Over the subsequent 25 years, a plethora of investigations have been undertaken to identify acidic proteins that are associated with biomineralized structures, and their functional role, but these have predominately been associated with shell formation (see e.g., Belcher et al., 1996; Falini et al., 1996; Weiss, et al., 2000, 2001; Pereira-Mouries et al., 2002; Gotliv et al., 2003). Indeed, mollusc shell matrix has been shown to be a complex mix of soluble and insoluble fractions, comprised of proteins, glycoproteins, proteoglycans and chitin (Marin & Luquet, 2004). However, due to the very low proportion of protein in the matrix of chiton teeth, the actual proteins present have not yet been identified and it is certainly an area that warrants further investigation, before we can fully understand the role of the matrix in chiton tooth biomineralization.

3.2 Matrix organization in chiton teeth

The matrix organization of chiton teeth has been the subject of investigation in a number of studies, each progressively using a more sophisticated suite of microscopical techniques. Evans et al. (1990, 1994) used light (LM), scanning electron (SEM) and transmission electron microscopy (TEM) to detail the complex arrangement of organic fibres in the teeth of *A. hirtosa* prior to the onset of mineralization. They described fibre density variation throughout the tooth cusp, which is matrix rich in the apatite region and matrix poor in the magnetite regions (Evans et al., 1990). At the posterior cutting surface of the tooth is a thin band of densely packed, fine fibres that appear stippled in TEM images, a layer that has also been observed in the cusps of *Plaxiphora albida* (Macey et al., 1994; Macey & Brooker, 1996). It is possible that this layer affords resistance to wear since it has been shown that it is highly resistant to chemical destruction. Adjacent to this layer in the cusps of both *A. hirtosa* and *P. albida* the fibres form into hollow tube-like structures, which become sparser towards the tooth core. In contrast, prominent long fibres running parallel to the tooth surface are featured in the anterior of the tooth. *In vitro* studies, resupplying iron-demineralized cusps with ferritin, showed alignment of the ferritin granules with the fibres, demonstrating the potential influence of the organics on initial mineral deposition in *in vivo* tooth cusps (Evans et al., 1994). In acid etched radula teeth of *Chiton olivaceus*, two types of organic tube-like structures were found to comprise the demineralized teeth (van der Wal et al., 1989), which the authors described as rod- and trough-shaped units. The presence of these units was later confirmed in the fully mineralized teeth of *A. echinata* using an environmental scanning electron microscope (ESEM) (Wealthall et al., 2005). These authors described the arrangement of the units in the teeth as having an overall 'fish scale'-like appearance (Figure 4). However, while van der Wal et al. (1989) identified a distinct discontinuity between the different mineralized regions of the tooth, Wealthall et al. (2005) showed the rod and trough units to be continuous throughout all biomineral regions of the tooth. In addition, there were no discrete borders between the regions and crystallites of each different mineral type were present in the adjacent mineral region. van der Wal et al. (1989) predicted that the alignment of the rod and trough units would afford the teeth a self-sharpening mechanism;

by channelling the direction of cracks in the teeth they controlled tooth wear, ensuring the cutting edge maintained an optimal chisel shape. Using more sophisticated visualization techniques, Wealthall et al. (2005) showed that cracks in the teeth were indeed propagated along the plane of the rod and trough units. A recent indentation fracture study determined that the organics play a fundamental role in the blunting of cracks and also their deflection at mineral interfaces (Weaver et al., 2010). Further elucidation of the fine structure of the organics in chiton teeth has been made possible through the utilization of a suite of microscopy equipment and techniques. Using a combination of focussed ion beam (FIB) section preparation, energy-filtered TEM (EFTEM) and scanning TEM (STEM) imaging, Saunders et al. (2011) were able to precisely target regions of interest in the radula teeth and investigate the relationships between organics and biominerals in the different mineral regions of the teeth and more specifically at their interfaces. They confirmed Wealthall et al's. (2005) finding that the organics are continuous between the mineral layers, and also determined that the original pre-mineralization fibre structure persists in the fully mineralized teeth. Using high-angle annular dark-field (HAADF) STEM, they revealed the true complexity of the organics, demonstrating the presence of individual fibres and bundles of fibres that aligned both along the transverse and the longitudinal axes of the cusps (Saunders et al., 2011).

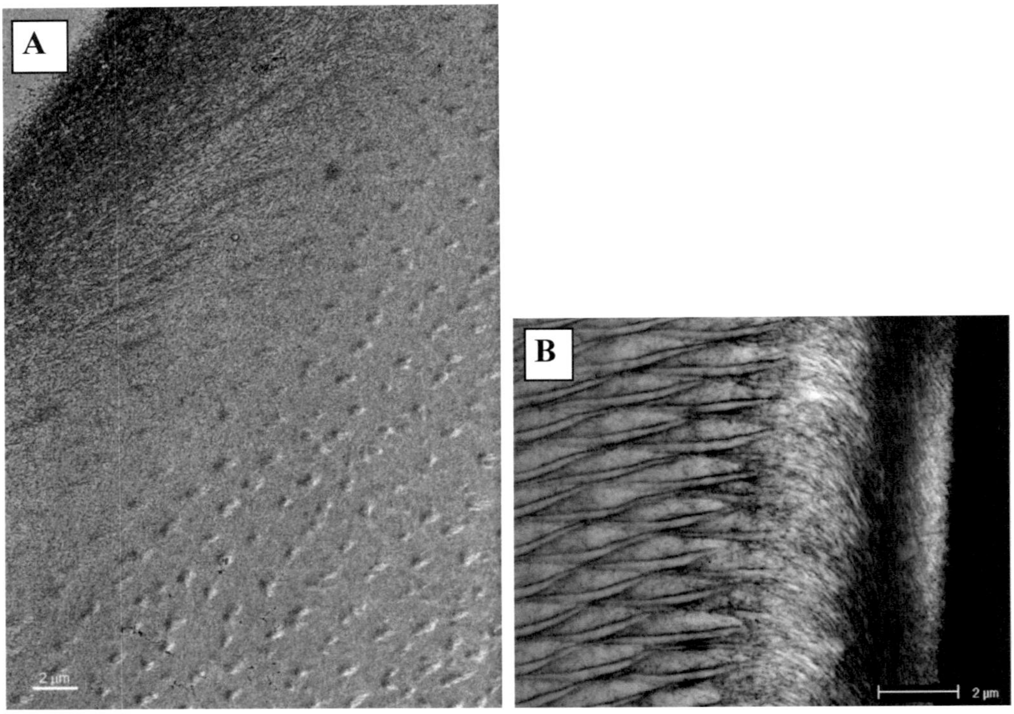

Fig. 4. Images of the organo-mineral interactions in the magnetite region of the tooth cusps of (A), *Acanthopleura hirtosa*, showing the 'leopard spot' appearance of the bundles of fibres, viewed using bright field TEM and (B), *Acanthopleura echinata,* showing the fish-scale appearance of the rod and trough structures, viewed with an environmental scanning electron microscope.

A re-examination of the dimensions of the organic structures and their arrangement in the tooth cusps of chitons, variously described as 'hollow tubes' (Evans et al., 1990), 'rods and troughs' (van der Wal et al., 1989), 'fish scales' (Wealthall et al., 2005) and 'leopard spots' (Saunders et al., 2011) (Figure 4), reveals that these authors are all describing the same structures, seen through the varying perspective of the techniques used. It is clear from all of the studies undertaken that the spatial distribution of the various minerals in the chiton tooth cusp is not a function of the physical organic structure, since this has been shown to be continuous from one region to the next (Wealthall et al., 2005; Saunders et al., 2011). As such, it is most likely that the mineral distribution is attributable to temporal changes in the chemical environment within the tooth at the different stages of tooth development. However, the studies indicate that the distribution and arrangement of the individual fibres and organic structures impact on the durability and structural integrity of the teeth, either through inhibition of cracks or the propagation of cracks along defined planes to optimize the teeth as a feeding tool. Nanoindentation studies of the cusps of *Cryptochiton stelleri* demonstrated the magnetite region to be harder than any previously reported biomineral structure, and, that the hardness was not affected by removal of the organic component of the cusp (Weaver et al., 2010). Due to the different mineral and organic composition of the teeth of *Cr. stelleri* to those of other chitons studied, such as *A. hirtosa*, further detailed physical studies need to be undertaken.

While there have been many studies that have visualized the organic fibres in the cusps, until recently, none have been able to examine the fibre composition and interaction with the mineral *in situ*. This is due mainly to the challenge of analyzing such limited quantities of nanoscale fibres that are buried deep within the minerals. Recently, Gordon & Joester (2011) utilized FIB processing and a pulsed-laser atom-probe to analyse 5-10nm fibres within the magnetite mineralized region of the teeth of *Chaetopleura apiculata*. They found that the fibres co-localized with either sodium or magnesium, and produced three-dimensional maps depicting the clustering of these cations with discrete bundles of fibres. The discovery of varying composition of individual fibres on the nanoscale has significant implications for our understanding of the functional roles of these fibres in the biomineralization process and deserves further investigation.

3.3 Mineral composition of chiton teeth

Sollas (1907) was the first to identify minerals in molluscan teeth, reporting the presence of silica and ferric oxides in the radula of limpets and chitons, respectively. Jones et al. (1935) confirmed the presence of ferric oxide in the radula, while Tomlinson (1959) noted that the major lateral teeth were actually black in color and reported that the chiton radula possessed magnetic properties. However, it was Lowenstam (1962) who determined that this property was attributable to the specific iron oxide magnetite (Fe_3O_4). This discovery prompted numerous studies over the subsequent 50 years investigating the biominerals in the major lateral teeth of chitons, which have shown that, while the iron oxide magnetite is ubiquitous to all chitons whose radulae have been described to date, there is a variety of other iron and calcium minerals that, while common to particular groups, are not universal to the class Polyplacophora.

The hard magnetite cap covering the cutting surface of chiton teeth has been estimated to have a Mohs hardness scale of 6 (Lowenstam, 1962), and the teeth have been reported to be the hardest biomineral structures known, exhibiting three times the hardness of human teeth or mollusc shell (Weaver et al., 2010). The iron mineral layers overlie a much softer central core (Lowenstam & Weiner, 1989), a design feature which has been suggested to impart significant shock absorbing capacity to the teeth (van der Wal et al., 2000; Shaw et al., 2009a).

While magnetite is found in the tooth cusps of all chitons, its physical distribution is genus specific. For example, in *Chiton* and *Acanthopleura* species, magnetite covers virtually the entire posterior surface, with the exception of a narrow band just superior to the junction between the tooth cusp and stylus, and continues over the distal tip, forming a narrow band on the anterior surface, which extends into a distinctive 'V-shaped' tab in the centre of the tooth (Figure 5) (Lowenstam, 1967; Lowenstam & Weiner, 1985; Brooker & Macey, 2001; Brooker et al., 2001, 2003; Shaw et al., 2008b; Saunders, et al., 2011). However, in *Cr. stelleri, Cryptoplax striata* and *Ch. apiculata* magnetite covers the entire anterior and posterior surface (Lowenstam & Weiner, 1985; Macey & Brooker, 1996; Gordon & Joester, 2011, respectively), while in *P. albida* it covers the posterior cusp surface and all but a small window at the base of the anterior cusp surface (Macey & Brooker, 1996). In addition to magnetite, a range of other iron oxides have been identified in the teeth.

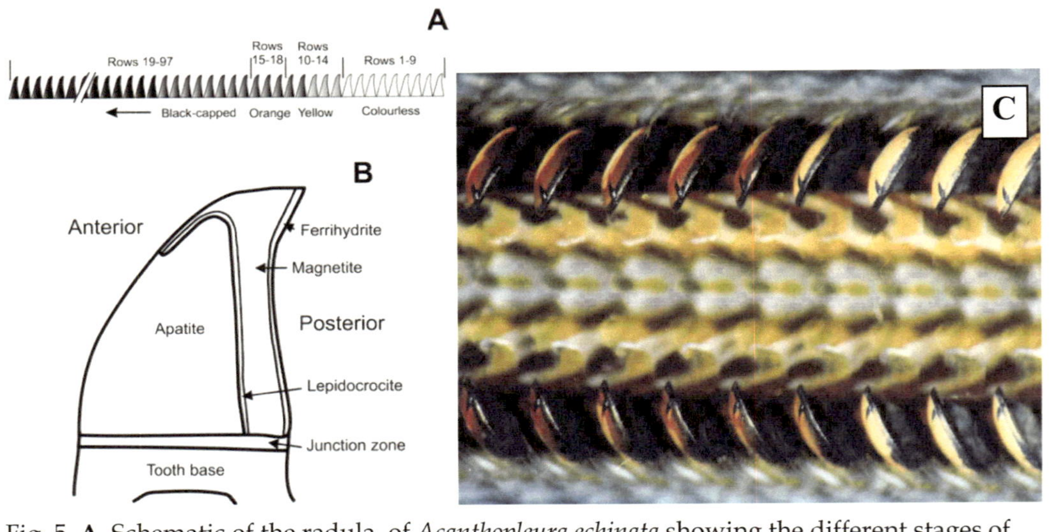

Fig. 5. **A**, Schematic of the radula, of *Acanthopleura echinata* showing the different stages of development, visible due to color differentiation. **B,** Diagram depicting the various regions in the tooth cusp. **C**, Light micrograph of the radula in the region of early onset of iron mineralization. The posterior, cutting surface of the prominent major lateral teeth is glossy black due to mineralization with magnetite. The red/brown color on the anterior surface of the teeth on the left half of the image is due to the presence of the iron oxide lepidocrocite, while the pale region on the anterior surface is yet to be infilled with hydroxyapatite.

Following the construction of the organic framework of the teeth, Lowenstam & Weiner (1989) described four stages of tooth mineralization in the Chitonida, the deposition of a transient iron mineral, its conversion to other iron minerals and the final infilling of the tooth core with an apatite mineral, with each mineral located in its own architecturally discrete compartment. The teeth in most developmental stages are easily discerned in the intact radula as they display distinctive colors (Figure 5A). At the immature end, the chitinous teeth are colorless and clear, moving towards the mature end of the radula are two or three rows of yellow teeth followed by a further two or three rows of red/brown teeth, which signal the deposition of ferrihydrite onto the matrix. Further on from these, magnetite is deposited and the teeth are glossy black from there to the anterior end of the radula. The deposition of the iron oxyhydroxide layer adjacent to the magnetite can be seen as a red/brown band on the anterior tooth surface (Figure 5C). However, it is more difficult to determine precisely where the tooth core becomes mineralized, although with careful observation this can be determined as when the core seen through the anterior surface changes from a translucent to an opaque appearance.

The first mineral deposited onto the organic matrix is the iron oxyhydroxide, ferrihydrite (Towe & Lowenstam, 1967; Kirschvink and Lowenstam, 1979; Kim et al., 1986a, 1989; van der Wal, 1990; Saunders et al., 2011), which persists through to the working teeth as a very thin layer over the posterior surface (Kirschvink & Lowenstam, 1979; Brooker, et al., 2003; Lee et al., 2003a). This initial mineral is subsequently converted to a range of other iron oxyhydroxide and iron oxide phases (Lowenstam & Weiner, 1989; Saunders et al., 2011). Lowenstam (1967) reported the orange/red material in the mature major lateral teeth of three chiton species: *C. tuberculatus; A. echinatum* (synonomous with *A. echinata*); and *A. spiniger* (synonomous with *A. gemmata*), to be the iron mineral lepidocrocite (γ-FeO.OH). Lepidocrocite was subsequently determined to be the mineral that abuts the magnetite layer in numerous chiton species (Lowenstam, 1967; Evans et al., 1994; Lee et al., 1998; 2000, 2003a; Brooker et al., 2003, 2006). While in Raman studies of the cusps of four chiton species, *P. albida* (Lee et al., 2003a), *A. rehderi, A. curtisiana* and *Onithochiton quercinus* (Lee et al., 2003b), a new iron biomineral, limonite, was reported in the region usually occupied by lepidocrocite in other *Acanthopleura* species. However, many of the earlier studies of the iron minerals used bulk sample analyses, predominantly XRD, and even the Raman analysis of Lee et al. (1998) had a limited spatial resolution of 10 - 15 μm.

Using a combination of FIB processing and TEM analysis, Saunders et al. (2009) were able to achieve vastly improved resolution over preceding studies, and demonstrated a far more complex arrangement of minerals than previously suggested. They differentiated two mineral phases in the iron oxyhydroxide layer of the cusps of *A. hirtosa*, the majority consisting of goethite (α-FeOOH) adjacent to the magnetite, with just a thin layer of lepidocrocite (γ-FeOOH) interior to this. This confirmed a much earlier finding of Kim et al. (1989) who observed needle-like crystals, identified by XRD as goethite, in the posterior region of *A. hirtosa* tooth cusps. The complexity of the mineral relationships in the cusps of chitons is further evidenced by the discovery that the minerals do not reside in separate compartments as originally proposed (Lowenstam, 1967; Lowenstam & Weiner, 1985). Rather, the indiscrete nature of mineral zone borders has been demonstrated with crystallites of mineral phases extending well into adjacent mineral zones (Wealthall et al., 2005, Saunders et al., 2009).

The core of the chiton tooth is the last region to be infilled and, as for the iron mineralized layers, a large variety of minerals have been identified in this region, and a complex pattern of deposition described. For a number of years it was believed that the core of chiton teeth consisted of one of two minerals; either an amorphous iron phosphate hydrogel, as reported in the tooth cores of *Cr. stelleri* (Lowenstam, 1972) and *Cry. striata*

(Macey et al., 1994), or an apatitic calcium phosphate (Lowenstam & Weiner, 1989; Evans & Alvarez 1999; Lee et al., 2000; Saunders et al., 2009). In a systematic study of the genus *Acanthopleura,* Brooker & Macey (2001) suggested that the minerals in the core of chiton teeth could be used as a taxonomic tool. They used EDS to determine the relative elemental composition of the tooth core of 18 *Acanthopleura* species, along with *Ischnochiton australis, P. albida* and *O. quercinus,* reporting significant differences in the percent weight composition of iron, phosphorous, calcium and magnesium. The core of *P. albida* teeth contained significant amounts of iron, and a lesser but substantial phosphorous component (Brooker & Macey, 2001). Lee et al. (2003a) later reported the core to be comprised of both limonite and lepidocrocite, but unable to find any evidence of an iron phosphate mineral, they surmised that the phosphate may be adsorbed onto the surface of the iron oxide minerals. The tooth core composition of *I. australis* is also unusual in that it contains small amounts of iron and magnesium in combination with almost equal proportions of calcium and phosphorus (Brooker et al., 2006), leading the authors to suggest that it may be comprised of whitlockite, $Ca_9(Mg,Fe)(PO_4)_6[PO_3(OH)]$, a mineral that has been identified in mineralized structures of a number of other invertebrate species (Lowenstam, 1972). However, further studies need to be undertaken to confirm this. In the chitons which possess an apatitic core, this mineral is preceded by deposition of an amorphous calcium phosphate (ACP) mineral, which has been shown to be either transformed into a crystalline apatite mineral (Lowenstam & Weiner, 1985), or to persist, providing the crystallographic conditions necessary for bonding of the iron and calcium phases (Saunders et al., 2009). The apatitic mineral has been variously reported as: francolite, a carbonated fluorapatite in *A. echinatum* (Lowenstam, 1967); dahllite, a carbonated hydroxyapatite in *A. haddoni* (Lowenstam & Weiner, 1985); or a carbonate and fluoride substituted apatite in *C. pelliserpentis* (Evans & Alvarez, 1999) and *A. hirtosa* (Evans et al., 1992).

3.4 Cellular role in chiton tooth biomineralization

The iron required for biomineralization of the chiton teeth originates as ferritin in the haemolymph (Kim et al., 1986b) and is delivered to the superior epithelial cells of the radula sac via the dorsal sinus (Nesson & Lowenstam, 1985; Shaw et al., 2009a). The radula tooth cusps of chitons are surrounded by the superior epithelium, with cells abutting all surfaces of the cusps. In addition, cells penetrate a pore in the tooth stylus, extending up the stylus canal to within 25 µm of within the junction of stylus and cusp (Nesson & Lowenstam, 1985; Shaw, et al., 2009b). Once the organic scaffolding of the radula teeth has been formed, the onset of mineralization is extremely rapid, and each of the stages in tooth mineralization, representing deposition of different minerals, has been shown to be very precisely controlled, and highly consistent within a species, with no more than a single row of variation between individuals (Shaw et al., 2009b). The first site of ion deposition is the junction between the tooth base and the cusp (Macey & Brooker, 1996, Shaw et al., 2009b) and it is believed that this region acts as a repository for ions that will subsequently migrate to the mineralizing front (Brooker et al., 2003, 2006; Sone et al., 2007; Shaw et al., 2009a).

Shaw et al. (2009a, 2009b) have described the ultrastructure of the cells of the superior epithelium and the stylus canal in the major lateral teeth of *A. hirtosa*. Both tissues comprise large columnar cells abundant with organelles and with a prominent basal nucleus. The number of mitochondria increases significantly just two rows prior to the onset of mineralization and they aggregate at the apical pole, close to the microvilli that abut the tooth (Figure 6). Likewise, the microvilli, which are poorly formed just a few rows prior to the onset of mineralization, dramatically increase in both abundance and length, extending up to 8 μm into the cytoplasm by the first orange colored tooth, which signals the appearance of ferrihydrite in the tooth. The cells of both the superior epithelium and the stylus canal are also primed and rich with ferritin granules, which start to proliferate at least five to six rows prior to the first orange tooth (Shaw et al., 2009a). While the form of iron as it passes from the cells to the tooth remains unknown, 8 nm iron rich particles have been observed throughout the cytoplasm, and also detected within the microvilli (Shaw et al., 2009a). The observation of ferritin granules in the cells abutting both the posterior and anterior cusp surfaces, as well as the cells of the stylus canal provides evidence for the multi-front delivery of ions to the cusp proposed by Brooker et al. (2003).

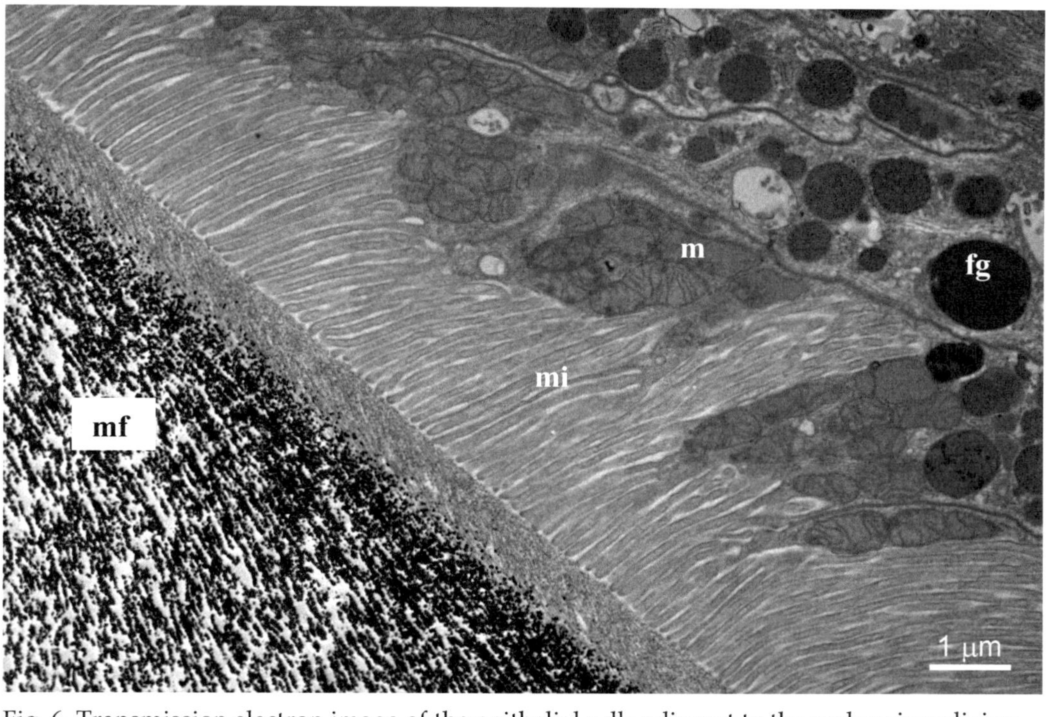

Fig. 6. Transmission electron image of the epithelial cells adjacent to the early mineralizing tooth, showing abundant mitochondria (m), electron dense ferritin granules (fg), microvilli (mi), and early mineral crystals attached to chitin fibres (mf) in the tooth cusp.

While TEM studies of cell ultrastructure at the earliest stages of mineralization have elucidated the process of cell development for iron delivery to the tooth cusps, iron is not the first element detected in chiton teeth. Prior to the appearance of ferrihydrite, sulphur ions are observed in the junction zone of *A. echinata* and possibly account for the yellow appearance of teeth several rows prior to the onset of mineralization (Macey & Brooker, 1996). This element is presumably associated with tanning of the organic matrix and coincides with the appearance of proteins in the matrix (Evans et al., 1991). Three rows prior to the first orange colored tooth, iron, phosphorus and calcium ions are detected at low levels in the junction zone with rapidly increasing concentrations over succeeding rows (Macey & Brooker, 1996). Shaw et al. (2009b) identified an internal pathway for channelling ions from the junction zone up through the tooth to the internal mineralizing surfaces. The formation of the 'plume' was found to coincide with the initial onset on mineralization. While it was originally assumed that the superior epithelial cells were the only ion delivery pathway for tooth mineralization (Nesson & Lowenstam, 1985; Kim et al., 1989), it seems logical that there would be other pathways. The initial deposition of ferrihydrite occurs over just a couple of rows and its mass conversion to magnetite happens in just one row. This rapid process would require an adequate supply of ions. Once magnetite is deposited, the mineral would form a physical barrier preventing the movement of ions through to the interior of the teeth, yet we know that other iron oxyhydroxides, and then calcium biominerals are deposited subsequent to and interior to the magnetite layer. Brooker et al. (2003) argued that it makes sense for ions to be delivered via a repository in the junction zone as well as via the, as yet, unmineralized anterior surface.

The cells of the superior epithelium are responsible for the control of the mineralization process and the delivery of ions to the tooth cusps during maturation (Nesson & Lowenstam, 1985; Shaw et al., 2008b, 2009a). In addition, cells within the hollow bases of the major lateral teeth, which are very similar in structure to the superior epithelium, are also responsible for delivering ions to internal mineralizing fronts within the cusp (Shaw et al., 2009b). Together, these tissues exert control over, and are responsible for, the complex sequence of biomineralization events that occur within the cusps, as well as the range of biominerals that are deposited there.

4. Methods and techniques

As the field of biomineralization inherently sits at the interface between materials science and biology it has often adopted methodology suited to both. This is mainly owing to the fact that researchers are required to look at the hard and soft tissues, both of which often require quite separate methodologies. Chiton teeth are no exception, and, due to the presence of magnetite, researchers have had to come up with creative methods for dealing with the practical challenges such hard structures present. The biggest issue that has confronted researchers in this field has been to reveal the biomineralization processes that are occurring *in situ*.

As described above, a range of experimental techniques have been developed and applied to investigations of chiton radula biomineralization, but the approach has evolved over the past 50 years, as more sophisticated techniques have become available. A summary of the change in emphasis over this time period is presented in Table 1. Initial studies used traditional chemical analysis to identify organic and inorganic components of the radula teeth. Subsequently, the organic matrix of unmineralized teeth was examined using light, scanning and transmission electron microscopy (SEM & TEM). While the mineral composition, crystallization and iron oxide phases of fully mineralized teeth have been analyzed using PIXIE analysis, X-ray diffraction, energy dispersive spectroscopy and Raman spectroscopy. More recently, focused ion beam (FIB) milling has been utilized to prepare precisely oriented sections in fully mineralized teeth (Saunders et al., 2009, 2011), and a combination of techniques have been used to obtain simultaneous information on the organic and mineral phases of fully mineralized chiton teeth: charge contrast imaging (Stockdale et al., 2009);

energy-filtered TEM to highlight the interactions between various mineral phases; and high-angle annular dark-field scanning TEM to demonstrate the continued existence of the organic matrix in fully mineralized teeth (Saunders et al., 2010, 2011).

Decade	Technique	Main findings
1960's - 70's	X-ray powder diffraction (XRD) Light microscopy (LM)	Physical properties Radula structure Bulk mineral composition Identification of: magnetite, lepidocrocite, goethite, apatite, Amorphous precursor phases
1980's	XRD, LM Transmission electron microscopy (TEM) Infrared spectroscopy (IR) Proton induced x-ray emission (PIXE) Proton induced γ-ray emission (PIGME)	Ferritin – source of iron Cusp epithelium Gross mineral architecture Identification of: dahllite Matrix control of mineralization
1990's	Scanning electron microscopy (SEM) Energy dispersive spectroscopy (EDS) Raman Atomic force microscopy (AFM)	Micro-scale mineral composition Mechanical properties
2000's	SEM, EDS, Raman Environmental SEM (ESEM) Energy filtered TEM (EFTEM) Electron microprobe analysis (EMPA)	Micro/nano-scale mineral composition Tooth development
2010 and beyond	TEM (EFTEM), SEM Focused ion beam processing (FIB) High-angle annular dark-field scanning TEM (HAADF-STEM) Atom Probe, micro/nano-CT, nano-indentation, molecular studies	Nano-scale mineral composition Tooth structure and mechanical properties. 3D structure Molecular basis of biomineralization.

Table 1. Summary of the evolution of techniques used in chiton biomineralization studies and the major findings from 1960 to the current day.

5. Future research

A recurring problem associated with the study of chiton teeth is the sheer morphological and functional complexity of the radula. In particular, the shape and positioning of the major lateral teeth is known to relate to how they are used for feeding (Shaw et al., 2010). Not only is this gross structure of importance from a functional context, but the overall functional morphology of the teeth no doubt has a strong bearing on the tooth's internal fine structure. Understanding this is especially important for studies aimed at elucidating the structural and mechanical properties of the teeth, where orientation can play a large role in the collection and interpretation of data. Future studies, should take a holistic approach, where the radula and teeth are observed across a range of spatial scales and placed in their functional context.

With the recent adoption by researchers in the field of chiton radula biomineralization of a range of newly evolved, sophisticated techniques, and still more that are in their infancy of application to this research, there are many exciting avenues to investigate in the coming years, some of which will be explored here.

5.1 The illusive organic matrix

A number of studies over the past 30 years have examined the organic matrix of radula teeth from the micro to the nano level, mostly concentrating on elucidating structural information. While we are starting to gain an appreciation of the intricate relationships between the organic and mineral phases in the cusp, we are still at a loss to understand precisely what we are looking at. The studies have depicted very complex arrangements of fibres in the tooth cups (van der Wal et al., 1989; Wealthall, et al., 2005; Saunders et al., 2011), but have revealed very little about the overall arrangement of fibres in the various regions of the cusp. We are continually hampered by this lack of perspective when it comes to determining orientations and interpreting our two dimensional electron micrographs (SEM and TEM). In order to reveal the three dimensional blueprint of the whole organic matrix, techniques such as micro- and nano-computed tomography, or even confocal microscopy, could be applied to these structures: only when we understand the bigger picture will we have a chance of correctly interpreting the finer detail we see at the nano-level.

Despite our acceptance that the organic matrix plays a fundamental role in mediating biomineralization in chiton teeth, we are still unsure about the extent of its role. If the arrangement of the matrix is consistent across widely varying mineral phases (Wealthall et al., 2005; Saunders et al., 2011), what controls the deposition of specific minerals in defined regions of the cusp? Is it a result of temporal changes in the chemical environment supplied by the superior epithelial cells at different stages of tooth development? Is it a result of small variations in the chemical composition of the matrix fibres, as suggested by Gordon & Joester (2011)? Or is it a combination of both or an, as yet, undiscovered feature of the matrix? In the past, our understanding of biomineralization process *in situ* has been repressed by issues of ultrathin section preparation across the layers of varying hardness in chiton teeth, but the recently applied technique of FIB, combined with advanced analytical microscopy, offers much opportunity to enhance our future understanding of matrix structure and interaction with the minerals.

In order to understand how the matrix directs biomineralization it is imperative that we determine the composition of the organic component of chiton teeth. It was not until more was understood about the nature of the proteins in mollusc shell, that models were able to be proposed to explain the properties of shell and the ways in which the proteins control the biomineralization processes in these structures (see for example: Levi-Kalisman et al., 2001). Even though proteins constitute a small proportion of the organics of chiton teeth, it is highly likely that they are imperative for initial mineral nucleation. With the complexity of mineral phases found in these structures, fine protein variations could well play a role in determining their spatial and temporal deposition along the chiton radula.

5.2 Molecular control of radula biomineralization processes

A new generation of molecular technologies, such as, genomics, transcriptomics and proteomics, is enabling a holistic approach to an investigation of the molecular basis of biomineralization. Over the past decade much has been achieved in progressing our understanding of the molecular control of molluscan shell biomineralization, with the characterization of proteins that promote or inhibit biomineralization or selectively enhance the formation of particular mineral phases (for example: lustrin A (Shen et al., 1997); mucoperlin (Marin et al., 2000); perlucin and perlustrin (Weiss et al., 2000); MS17 (Zhang et al., 2003); Prismalin-14 (Suzuki et al., 2004)). The successful knockdown of a biomineralization gene like Pif (Suzuki et al., 2009) facilitating demonstration of the function of the gene *in situ*, holds immense promise for enhancing our understanding of the molecular control of biomineralization in the future. However, an understanding of the molecular control of chiton radula mineralization is lagging far behind; there is currently, no published information available concerning the molecular basis of biomineralization in the chiton radula. Specifically, there is no knowledge of the nature and composition of the proteins that regulate and/or constitute the radula matrix, or the genes that encode them. Adopting molecular techniques has the potential to answer fundamental questions in regard to radula biomineralization. We have already commenced some preliminary work in this area, utilizing both a microarray and transcriptomics approach in the creation of a radula specific cDNA library. Differentially expressed candidate genes have been identified that code for a range of putative protein domains showing homology to proteins with confirmed relevance to the biomineralization process (unpublished data). While subsets of these genes have been specifically associated with the different development stages of the chiton radula (pre-mineralizing, iron deposition and calcium deposition regions), there is still a long way to go in determining the actual function of the genes in radula biomineralization. The biggest hurdle in a molecular approach to understanding radula biomineralization is the severe lack of molecular data on any aspect of chiton physiology. However, there is potential for leaps forward in this area. For instance, if critical biomineralization genes can be identified and knocked down, or manipulated through the use of innovative protocols for the cultivation of chitons such as iron limited conditions (Shaw et al., 2007), which delay the onset of iron mineralization in the radula, facilitating molecular studies of up and down regulation of genes by ferritin in the radula epithelium.

5.3 Varying biomineralization strategies employed by chitons

Some studies have undertaken a broad sweep approach to confirm whether radula biomineralization in chitons is consistent, such as the confirmation of magnetite in the cusps of all species, or of lepidocrocite across a range of species. However, the majority of studies have concentrated on a limited number of species, with most focusing on *A. hirtosa*. Lowenstam & Weiner (1989) differentiated between two main strategies in *Cryptochiton* and the Chitonida based on the mineral composition of the core. However, it is now clear that the suite of iron and calcium minerals that chitons are able to incorporate into their teeth is more complex than originally described. Brooker et al. (2006) proposed a possible phylogenetic basis to the variety of minerals in radula cusps of different chiton taxa, and it would be fruitful to delve deeper into the minerals deposited by *Ischnochiton* and *Plaxiphora* for example, which do not appear to fit either of Lowenstam & Weiner's (1985) models. It would be feasible to apply molecular techniques to this project to determine if the different biomineralization models reflect varying genetic profiles.

While quite different mineral and organic compositions have been identified in the teeth of many chitons, for example *Cry. striata* and *A. hirtosa*, this has not been followed up with further detailed physical studies. Nanoindentation studies, like that undertaken on *Cr. stelleri,* need to be extended to other species to gain a fuller understanding of the role of the organics and minerals and the symbiotic relationships between them.

5.4 The interfaces between mineral phases

While the various mineral phases and their architecture have been well described for the radula teeth of some chiton genera, the interfaces connecting these mineral phases are less well understood, yet make a fundamental contribution to the materials properties of chiton teeth. For instance, the potential barrier to crack propagation afforded by an intermediate iron oxide layer between the magnetite and calcified regions of the cusp, or the persistence of an amorphous calcium phase between the intermediate iron oxide layer and the fully calcified core of mature teeth. This mechanism for bonding crystallographically disparate materials together further enhances the structural integrity of this multi-phase biomineral structure. While the chiton has adopted this strategy through evolutionary trial and error, it is a stunning feat when considered from a synthetic materials perspective, and there is clearly a great deal that can be learnt from further studies of these fine-scale mineral interactions.

6. Conclusion

Over the past 50 years, there has been extensive investigation into the processes of biomineralization in the chiton radula. With its spatial and temporal presentation of varied and complex organo-mineral interactions there is an immense amount to unravel and, judging by the increasing number of questions the research poses, we have barely scratched the surface. It is indeed a unique tissue for an investigation of biomineralization processes, the understanding of which holds a distinctive promise for applications to fields such as medicine, engineering, materials science, nanotechnology and biomimetics.

7. References

Addadi, L. & Weiner, S. (1985). Interactions between acidic proteins and crystals: Stereochemical requirements in biomineralization. *Proceedings of the National Academy of Sciences of the United States of America,* Vol. 82, No. 12, pp. 4110-4114.

Belcher, A.M., Wu, X.H., Christensen, R.J., Hansma, P.K., Stucky, G.D. & Morse, D.E. (1996). Control of crystal phase switching and orientation by soluble mollusc-shell proteins. *Nature,* Vol. 381, No. 6577, pp. 56-58.

Birchall, J.D. (1989). The importance of the study of biominerals to materials technology. In: *Biomineralization: Chemical and Biochemical Perspectives,* S.Mann, J. Webb, & R.J.P. Williams, (Eds.), 491–509, VCH Verlagsgesellschaft, Weinheim.

Brooker, L.R., & Macey, D.J. (2001). Biomineralization in chiton teeth and its usefulness as a taxonomic character in the genus *Acanthopleura* Guilding, 1829 (Mollusca: Polyplacophora). *American Malacological Bulletin,* Vol. 16, pp. 203–215.

Brooker, L.R., Lee, A.P., Macey, D.J. & Webb, J. (2001). Molluscan and other marine teeth. In: *Encyclopaedia of Materials: Science and Technology.* R.W. Cahn & P.D. Calvert (Eds.) 5186-5189, Elsevier Science Ltd., Oxford.

Brooker, L.R., Lee, A.P., Macey, D.J., van Bronswijk, W., & Webb, J. (2003). Multiple-front iron-mineralisation in chiton teeth (*Acanthopleura echinata*: Mollusca: Polyplacophora). *Marine Biology*, Vol. 142, pp. 447–454.

Brooker, L. R. (2004). Revision of *Acanthopleura* Guilding, 1829 (Mollusca: Polyplacophora) based on light and electron microscopy. PhD dissertation, Murdoch University, Murdoch. Australasian Digital Thesis Program, Retrieved from http://researchrepository.murdoch.edu.au/479/.

Brooker, L. R., Lee, A. P., Macey, D.J., Webb, J. & van Bronswijk, W. (2006). *In situ* studies of biomineral deposition in the radula teeth of chitons of the suborder Chitonina. *Venus*, Vol. 65, No. 1-2, pp. 71-80.

Eernisse, D.J. & Kerth, K. (1988). The initial stages of radular development in chitons (Mollusca: Polyplacophora). *Malacologia*, Vol. 28, No. 1-2, pp. 95-103.

Eernisse, D.J. & Reynolds, P.D. (1994). Polyplacophora, In: *Microscopic Anatomy of Invertebrates Volume 5: Mollusca I*, F.W. Harrison & A.J. Kohn (Eds.), 55-110, Wiley-Liss Inc, New York.

Evans, L.A., Macey, D.J. & Webb, J. (1990). Characterization and structural organization of the organic matrix of radula teeth of the chiton *Acanthapleura hirtosa*. *Philosophical Transactions of the Royal Society of London, B*, Vol. 329, pp. 87–96.

Evans, L.A., Macey, D.J. & Webb, J. (1991). Distribution and composition of the matrix protein in the radula teeth of the chiton *Acanthopleura hirtosa*. *Marine Biology*, Vol. 109, pp. 281-286.

Evans, L.A., Macey, D.J. & Webb, J. (1992). Calcium biomineralization in the radula teeth of the chiton, *Acanthopleura hirtosa*. *Calcified Tissue International*, Vol. 51. pp. 78-82.

Evans, L.A., Macey, D.J. & Webb, J. (1994). Matrix heterogeneity in the radular teeth of the chiton *Acanthopleura hirtosa*. *Acta Zoologica*, Vol.75, pp. 75–79.

Evans, L. A. & Alvarez, R. (1999). Characterization of the calcium biomineral in the radular teeth of *Chiton pelliserpentis*. *Journal of Biological Inorganic Chemistry* Vol. 4, No. 2, pp. 166-170.

Falini, G., Albeck, S., Weiner, S. & Addadi, L., (1996). Control of aragonite or calcite polymorphism by mollusk shell macromolecules. *Science*, Vol. 271, pp. 67–69.

Fretter, V. & Graham, A. (1962). *British prosobranch molluscs, their functional anatomy and ecology*. Ray Society, London.

Gordon, L. & Joester, D. (2011). Nanoscale chemical tomography of buried organic–inorganic interfaces in the chiton tooth, *Nature*, Vol. 469, pp. 194-198.

Gotliv, B.A., Addadi, L. & Weiner, S., (2003). Mollusk shell acidic proteins: in search of individual functions. *Chembiochem*, Vol. 4, pp. 522–529.

Graham, A. (1973). The anatomical basis of function in the buccal mass of Prosobranch and Amphineuran molluscs, *Journal of Zoology, London*, Vol. 169, pp.317-348.

Jones, E.I., McCance, R.A. & Shackleton, L.R.B. (1935). The role of iron and silica in the structure of the radular teeth of certain marine molluscs. *Journal of Experimental Biology*, Vol, 12, pp.59–64.

Kaas, P. and Jones, A.M. (1998). Class Polyplacophora: Morphology and Physiology, In: *Mollusca: The Southern Synthesis Part A, Fauna of Australia*, P.L. Beesley, G.J.B. Ross & A.Wells (Eds.), 163-174, CSIRO, Melbourne.

Kim, K.S., Webb, J., Macey, D.J. & Cohen, D.D. (1986a). Compositional changes during biomineralization of the radula of the chiton *Clavarizona hirtosa. Journal of Inorganic Biochemistry*, Vol. 28, pp. 337-345.

Kim, K.S., Webb, J. & Macey, D.J. (1986b). Properties and role of ferritin in the hemolymph of the chiton *Clavarizona hirtosa. Biochimica et Biophysica Acta*, Vol. 884, pp. 387-394.

Kim, K.S., Macey, D.J., Webb, J. & Mann, S. (1989.) Iron mineralisation in the radula teeth of the chiton *Acanthopleura hirtosa. Proceedings of the Royal Society of London* Series B, Vol. 237, pp. 335–346.

Kirschvink, J.L. & Lowenstam, H.A. (1979). Mineralization and magnetization of chiton teeth: paleomagnetic, sedimentalogic and biologic implications of organic magnetite. *Earth and Planetary Science Letters*, Vol. 44, pp. 193-204.

Lee, A.P., Webb, J., Macey, D.J., van Bronswijk, W., Savarese, A.R. & de Witt, G.C. (1998). *In situ* Raman spectroscopic studies of the teeth of the chiton *Acanthopleura hirtosa. Journal of Biological Inorganic Chemistry*, Vol. 3, pp. 614–619.

Lee, A.P., Brooker, L.R., Macey, D.J., van Bronswijk, W. & Webb, J. (2000). Apatite mineralization in teeth of the chiton *Acanthopleura echinata.* Calcified Tissue International, Vol. 67, pp. 408–415.

Lee, A.P., Brooker, L.R., Macey, D.J., Webb, J. & Van Bronswijk, W. (2003a). A new biomineral identified in the cores of teeth from the chiton *Plaxiphora albida. Journal of Biological Inorganic Chemistry*, Vol. 8, pp. 256-262.

Lee, A.P., Brooker, L.R., Van Bronswijk, W., Macey, D.J. & Webb, J. (2003b). Contribution of Raman spectroscopy to identification of biominerals present in teeth of *Acanthopleura rehderi, Acanthopleura curtisiana,* and *Onithochiton quercinus. Biopolymers*, Vol. 72, pp. 299–301.

Levi-Kalisman, Y., Falini, G., Addadi, L. and Weiner, S. (2001). Structure of the nacreous organic matrix of a bivalve mollusk shell examined in the hydrated state using Cryo-TEM. *Journal of Structural Biology*, Vol. 135, pp. 8-17.

Lowenstam, H.A. (1962). Magnetite in denticle capping in recent chitons (Polyplacophora). *Geological Society of America Bulletin*, Vol. 73, pp. 435–438.

Lowenstam, H. A. (1967). Lepidocrocite, an apatite mineral, and magnetite in teeth of chitons (Polyplacophora). *Science*, Vol. 56, pp. 1373-1375.

Lowenstam, H.A. (1972). Phosphatic hard tissues of marine invertebrates, their nature and mechanical function, and some fossil implications. *Chemical Geology*, Vol 9, pp. 153-166.

Lowenstam, H. A. & Weiner, S. (1985). Transformation of amorphous calcium phosphate to crystalline dahllite in the radula teeth of chitons. *Science*, Vol. 227, pp. 51-52.

Lowenstam, H.A., & Weiner, S. (1989). Mollusca, In: *On Biomineralization*, H.A. Lowenstam & S. Weiner (Eds.) 88-305 Oxford University Press, Oxford.

Macey, D.J., Webb, J. & Brooker, L.R. (1994). The structure and synthesis of biominerals in chiton teeth. *Bulletin de l'Institut oceanographique , Monaco,* Vol. 4, No. 1, pp. 191–197.

Macey, D.J. & Brooker, L.R. (1996). The junction zone: Initial site of mineralization in radula teeth of the chiton *Cryptoplax striata* (Mollusca: Polyplacophora. *Journal of Morphology*, Vol. 230, pp. 33–42.

Marin, F., Corstjens, P., de Gaulejac. B., de Vrind-De Jong, E. & Westbroek, P. (2000). Mucins and molluscan calcification molecular characterization of mucoperlin, a novel mucin-like protein from the nacreous shell layer of the fan mussel *Pinna nobilis*

(Bivalvia, Pteriomorphia), *The Journal of Biological Chemistry*, Vol. 275, No. 27, pp. 20667–20675.

Marin, F. & Luquet, G. (2004). Molluscan shell proteins. *C. R. Paleontology*, Vol. 3, pp. 469 - 492

Nesson, M.H. & Lowenstam, H.A. (1985). Biomineralization processes of the radula teeth of chitons. In: *Magnetite Biomineralization and Magnetoreception in Organisms*, J.L. Kirschvink, D.S. Jones, & B.J. MacFadden, (Eds.), pp. 333–361. Plenum Press, New York.

Pereira-Mouries, L., Almeida, M.J., Ribeiro, C., Peduzzi, J., Barthelemy, M., Milet, C. & Lopez, E. (2002). Soluble silk-like organic matrix in the nacreous layer of the bivalve *Pinctada maxima*. *European Journal of Biochemistry*, Vol. 269, pp. 4994–5003.

Runham, N.W. (1962). Rate of replacement of the molluscan radula. *Nature*, Vol. 194, pp. 992–993.

Saunders, M., Kong, C., Shaw, J.A., Macey, D.J. & Clode, P.L. (2009). Characterization of biominerals in the radula teeth of the chiton, *Acanthopleura hirtosa*. *Journal of Structural Biology*, Vol. 167, No. 1, pp. 55–61.

Saunders, M., Shaw, J.A., Clode, P.L., Kong, C. & Macey, D.J. (2010). Fine-scale analysis of biomineralized mollusc teeth using FIB and TEM. *Microscopy Today*, Vol. 18, No. 1, pp. 24–28.

Saunders, M., Kong, C., Shaw, J.A., & Clode, P.L. (2011). Matrix-mediated biomineralization in marine mollusks: A combined transmission electron microscopy and focused ion beam approach. *Microscopy and Microanalysis*. Vol. 17, pp. 220–225.

Shaw, J.A., Brooker, L.R. & Macey, D.J. (2002). Radular tooth turnover in the chiton *Acanthopleura hirtosa* (Blainville, (1825) Mollusca: Polyplacophora). *Molluscan Research*, Vol. 22, pp. 93–99.

Shaw, J.A., Macey, D.J., Brooker, L.R., Clode, P.L. & Stockdale, E.J. (2007). The stylus canal: a conduit for the delivery of ions to the mineralizing tooth cusps of chitons? *Proceedings of the 9th International Symposium on Biomineralization: Biomineralization from Paleontology to Materials Science*, J.L. Arias & M.S. Fernándes (Eds.), pp.187-192 Editorial Universitaria, Santiago, Chile.

Shaw, J.A., Macey, D.J. & Brooker, L.R. (2008a). Radula synthesis by three species of iron mineralizing molluscs: Production rate and elemental demand. *Journal of the Marine Biological Association of the UK*, Vol. 88, No. 03, pp. 597–601.

Shaw, J.A., Macey, D.J., Clode, P.L., Brooker, L.R., Webb, R.I., Stockdale, E.J. & Binks, R.M. (2008b). Methods of sample preparation of epithelial tissue in chitons (Mollusca: Polyplacophora). *American Malacological Bulletin*, Vol. 25, pp. 35–41.

Shaw, J.A., Macey, D.J., Brooker, L.R., Stockdale, E.J., Saunders, M. & Clode, P.L. (2009a). Ultrastructure of the epithelial cells associated with tooth biomineralization in the chiton *Acanthopleura hirtosa*. *Microscopy and Microanalysis*, Vol. 15, No. 2, pp. 154–165.

Shaw, J.A., Macey, D.J., Brooker, L.R., Stockdale, E.J., Saunders, M. & Clode, P.L. (2009b). The chiton stylus canal: An element delivery pathway for tooth cusp biomineralization. *Journal of Morphology*, Vol. 270, pp. 588-600.

Shaw, J.A., Macey, D.J., Brooker, L.R. & Clode, P.L. (2010). Tooth use and wear in three iron-mineralizing mollusc species. *Biological Bulletin*, Vol. 218, pp. 132-144.

Shen, X., Belcher, A. M., Hansma, P. K., Stucky, G. D. and Morse, D. E. (1997). Molecular cloning and characterization of lustrin A, a matrix protein from shell and pearl nacre of *Haliotis rufescens*. *Journal of Biological Chemistry*, Vol. 272, pp. 32472–32481.

Sollas, I. B. J. (1907). The molluscan radula: its chemical composition, and some points in its development. *Quarterly Journal of Microscopical Science*, Vol. 51, pp. 115-136.

Sone, E.D., Weiner, S. & Addadi, L. (2007). Biomineralization of limpet teeth: A cryo-TEM study of the organic matrix and the onset of mineral deposition. *Journal of Structural Biology*, Vol. 158, pp. 428–444.

Stockdale, E. J., Shaw, J.A., Macey, D.J. & Clode, P.L. (2009). Imaging organic and mineral phases in a bomineral using novel contrast techniques. *Scanning*, Vol. 31, pp. 11–18.

Suzuki, M., Murayama, E., Inoue, H., Ozaki, N., Tohse, H., Kogure, T & Nagasawa, H. (2004). Characterization of Prismalin-14, a novel matrix protein from the prismatic layer of the Japanese pearl oyster (*Pinctada fucata*). *Biochemical Journal*, Vol. 382, pp. 205–213.

Suzuki, M., Saruwatari, K., Kogure, T., Yamamoto, Y., Nishimura, T., Kato, T. & Nagasawa, H. (2009). An acidic matrix protein, Pif, is a key macromolecule for nacre formation. *Science*, Vol. 325, pp. 1388-1390.

Tomlinson, J. T. (1959). Magnetic properties of chiton radulae. *The Veliger*, Vol. 2, No. 2, p. 36.

Towe, K.M. & Lowenstam, H.A. (1967). Ultrastructure and development of iron mineralization in the radular teeth of *Cryptochiton stelleri* (Mollusca). *Journal of Ultrastructure Research*, Vol. 17, pp. 1–13.

van der Wal, P., Videler, J.J., Havinga, P. & Pel, R. (1989). Architecture and chemical composition of the magnetite-bearing layer in the radula teeth of *Chiton olivaceus* (Polyplacophora). In: *Origin, Evolution, and Modern Aspects of Biomineralization in Plants and Animals*, R.E. Crick (Ed.), pp. 153–166, Plenum Press, New York.

van der Wal, P. (1990). Structure and formation of the magnetite-bearing cap of the polyplacophoran tricuspid radula teeth. In: *Iron Biominerals*. R. B. Frankel & R. P. Blakemore (Eds.), pp. 221-229, Plenum Press, New York.

van der Wal, P., Giesen, H.J., Videler, J.J., (2000). Radular teeth as models for the improvement of industrial cutting devices. *Materials Science & Engineering C – Biomimetic and Supramolecular Systems*, Vol. 7, pp. 129–142.

Wealthall, R.J., Brooker, L.R., Macey, D.J. & Griffin, B.J., 2005. Fine structure of the mineralized teeth of the chiton *Acanthopleura echinata* (Mollusca: Polyplacophora). *Journal of Morphology*, Vol. 265, pp. 165–175.

Weaver, J.C., Wang, Q., Miserez, A., Tantuccio, A., Stromberg, R., Bozhilov, K.N., Maxwell, P., Nay, R., Heier, S.T., Di- Masi, E. & Kisailus, D. (2010). Analysis of an ultra hard magnetic biomineral in chiton radular teeth. *Materials Today*, Vol. 13, No. 1–2, pp. 42–52.

Weiss, I.M., Kaufmann, S., Mann, K. & Fritz, M. (2000). Purification and characterization of perlucin and perlustrin, two new proteins from the shell of the mollusc *Haliotis laevigata. Biochemical and Biophysical Research Communications,* Vol. 267, No. 1, pp. 17-21.

Weiss, I.M., Göhring, W., Fritz, M. & Mann, K. (2001). Perlustrin, a *Haliotis laevigata* (abalone) nacre protein, is homologous to the insulin-like growth factor binding protein N-terminal module of vertebrates, *Biochemical and Biophysical Research Communications,* Vol. 285, pp. 244 - 249.

Zhang, Y., Xie, L., Meng, Q., Jiang, T., Pu, R., Chen, L. and Zhang, R. (2003). A novel matrix protein participating in the nacre framework formation of pearl oyster, *Pinctada fucata. Comparative Biochemistry and Physiology, B,* Vol. 135, pp. 565–573.

5

Cartilage Calcification

Ermanno Bonucci[1] and Santiago Gomez[2]
[1]*La Sapienza University, Rome - Policlinico Umberto I*
Department of Experimental Medicine and Pathology, Roma
[2]*Department of Pathology, Medical School, University of Cádiz, Cádiz*
[1]*Italy*
[2]*Spain*

1. Introduction

Nineteenth century histologist Ranvier discovered the point of ossification and the calcification of cartilage as observed in embryos (Ranvier, 1875). By skillfully using simple methods such as hand-cutting with a razor, or with a lead-screw microtome of his own invention (still marketed as the Ranvier hand microtome), and chromic acid and carmine to stain, he observed the deposition of calcareous salts around cartilage capsules. Ranvier described chondrocytes and their distinctive arrangement in series to yield larger capsules. He observed how calcified capsules open into one another to form anfractuous cavities that become the earliest marrow spaces, and wondered what determines the resorption of the walls. After injecting Prussian blue to mark vessels in growing animals, he described wall destruction as proceeding selectively in the direction of vessel growth (Figure 1a). Soon after, Schäfer described how to prepare fresh or fixed cartilage sections, and recommended the use of osmic acid, silver nitrate and gold chloride (Schäfer, 1897). He then (Schäfer, 1907) published a series of four colour drawings comprising all the stages of ossification (Figure 1b); these drawings have been a source of inspiration for all later histology textbooks (see, for instance, Figure 1 c). Subsequent methodological advances in microscopy and microtomy allowed confirmation of all the early observations. In the first third of the twentieth century histology textbooks by Cajal, Bouin, Möllendorff, Maximow & Bloom, Cajal & Tello, Levi and Di Fiori, described endochondral ossification, the ossification centre, and the growth plate as we know them today. Interestingly, some doubts persisted about hypertrophic chondrocytes. Using silver reducing methods, Cajal and Tello described the well-developed Golgi apparatus of hypertrophic chondrocytes, and how it is reduced and fragmented in the last row of cells (Figure 1d), as well as the vacuolization of chondrocytes that makes these cells hypertrophic (Figure 1e).

The modern view has not changed, though many details have been added. Hypertrophic chondrocytes promote vascular invasion by producing a growth factor (Allerstorfer et al., 2010). They are no longer considered degenerating cells trapped within a calcified crust, but living, metabolically active cells (Farnum et al., 1990), as their environment is not hypoxic. The morphology and ultrastructure of hypertrophic chondrocytes are highly dependent on the methods used. With aqueous fixation procedures, chondrocytes appear shrunken. When the cartilage is processed by high-pressure freezing, freeze-substitution and low temperature embedding (Hunziker et al., 1984), followed by fixation in the presence of cationic dyes such as ruthenium hexammine trichloride, its hypertrophic chondrocytes retain a configuration comprising intact membranes attached to the pericellular rim, intact organelles and mitochondria. These studies demonstrate that chondrocytes are fragile cells, and that their participation in the mineralization process requires their functional viability.

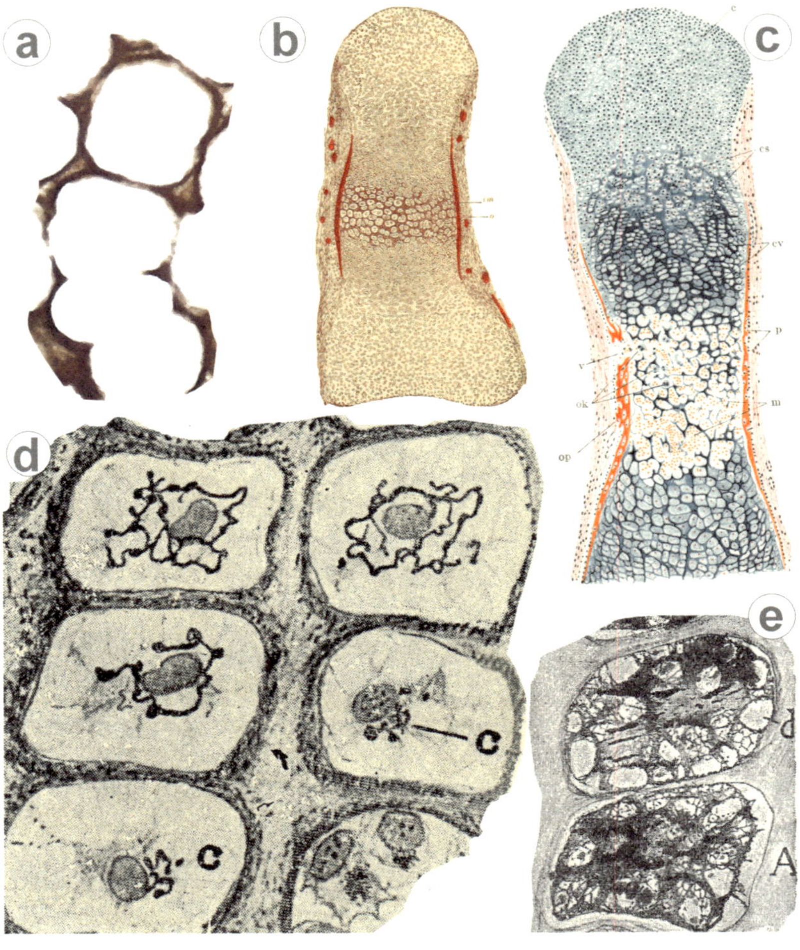

Fig. 1. (a) Ranvier's observation on the aperture of calcified capsules; (b) initial cartilage calcification in an embryo cartilage model stained with magenta spirit (from Schäfer); (c) primary ossification centre (from Möllendorff); (d) Golgi apparatus in hypertrophic chondrocytes (from Cajal & Tello); (e) vacuolization in hypertrophic chondrocytes (from Cajal & Tello).

The fate of hypertrophic chondrocytes is still a controversial issue. In many instances, especially in embryonic cartilage, they survive and dedifferentiate into osteoblasts (revised by Hall, 2005). However, other evolutions are possible, because the last row of cells is usually apoptotic. By eliminating cells, apoptosis establishes a dynamic equilibrium in the growth plate; it is not, anyway, considered to be directly involved in the mineralization process (Pourmand et al., 2007), but may interfere with it.

2. The chondrocyte event sequence for matrix calcification

Cartilage calcification is a regular and efficient process orchestrated by its chondrocytes, which go through a series of morphological/functional changes called the chondrocyte differentiation sequence, or, more properly, the mineralizing sequence. In histological sections treated with a general staining agent, such as toluidine blue, this sequence is easy to follow, since groups of chondrocytes show synchronous changes near the calcification zone (Figure 2a). The series comprises the proliferation, maturation, and hypertrophy zones. Staining sections with the von Kossa method allows the distribution of calcium deposits to be recognized (Figure 2b).

Mammalian and avian growth cartilages have been the subject of numerous studies as a model for cartilage calcification. Histologically, they are similar types of cartilage, but differences are found with respect to the mineralizing sequence. In the (mammalian) growth plate cartilage, the sequence is shorter, and calcium deposits are only seen around the last rows of hypertrophic chondrocytes (usually 2-3 cells); conversely, in avian cartilage scores of chondrocytes surrounded by calcified deposits are easy to find. In the former, apoptosis is observed in the last row of chondrocytes (Figure 2c), whereas, in the latter, apoptotic chondrocytes are not found within the sequence, and apoptosis only appears close to the resorption limits. As a result, the mineralizing sequence in the mammalian growth plate is shorter than in avian cartilage because it is abruptly interrupted by apoptosis.

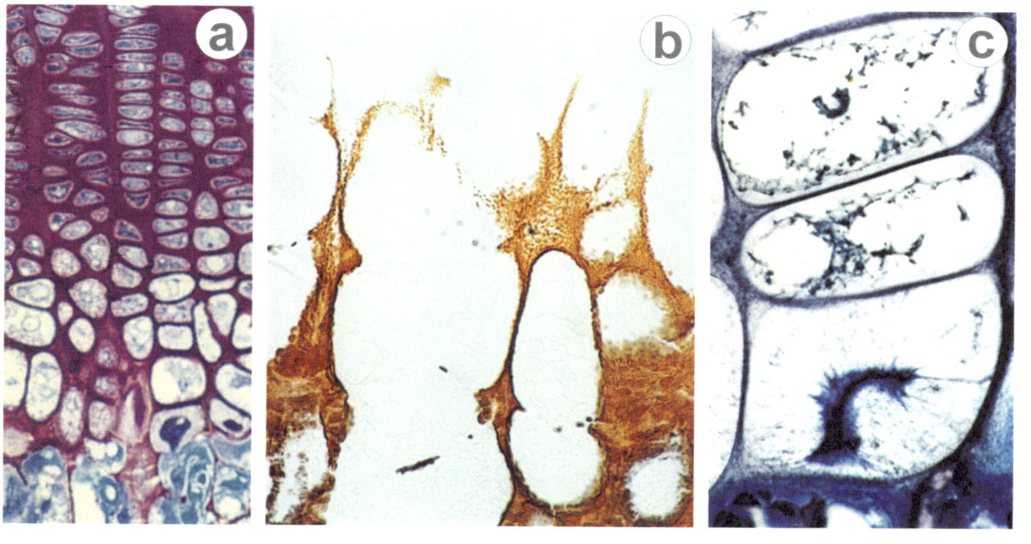

Fig. 2. Growth plate (a), mineral (b) and apoptosis in the last row of chondrocytes (c).

Whatever the type of cartilage, the best way of identifying the mineralizing potential of a given sequence is to stain its alkaline phosphatase and calcium-binding sites. Figure 3 shows the chondrocyte sequence in rat growth plate, as observed by LM methods for alkaline phosphatase and for calcium-binding sites. Figure 4 shows the results of the same LM methods, but applied to embryonic chick cartilage.

Staining the alkaline phosphatase (TNAP, tissue non-specific isoenzyme of alkaline phosphatase), using glycero-phosphate or azo-dye methods, determines whether a cartilage is entering calcification, and signals the beginning of the sequence. The chondrocytes show staining of the plasma-membranes and of a thin rim of adjacent matrix. In both types of cartilage, the early maturation chondrocytes and the matrix surrounding them are TNAP-positive (Figure 3a-c; Figure 4a). This matrix is not yet calcified, and TNAP staining becomes negative wherever the matrix is calcified (Figure 3c).

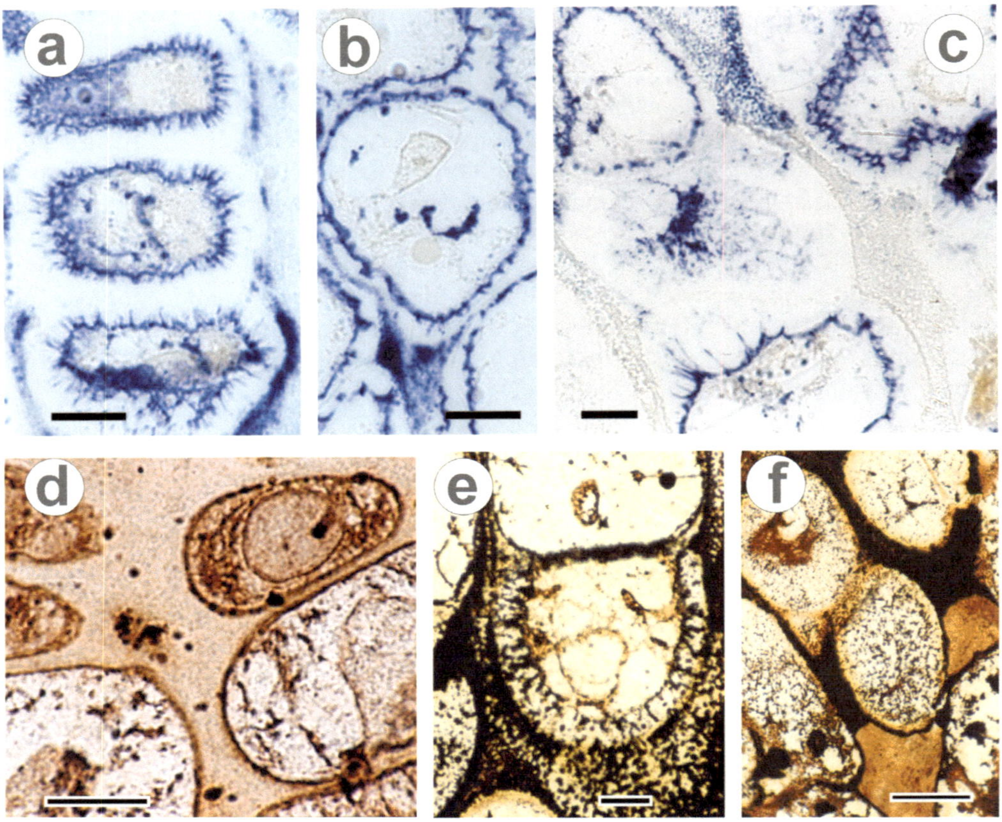

Fig. 3. TNAP (a-c); and calcium-binding staining (d-f) in rat growth plate. (Bar = 10 μm).

Ultrastructurally, plenty of matrix vesicles (MVs; Anderson, 1967, 1969; Bonucci, 1967, 1970) are found in these zones. These are the TNAP-rich MVs first described by Matsuzawa and Anderson (1971) and later confirmed for both types of cartilage (Akisaka & Gay, 1985; Bonucci et al., 1992; Takagi & Toda, 1979; Takechi & Itakura, 1995 a, 1995b). When these TNAP-rich MVs are isolated and cultured in mineralizing solutions, they show mainly extravesicular apatite deposition (Boskey et al., 1994). They are considered to arise mostly from maturation and early hypertrophic cells (Anderson, 1995). Hypertrophic chondrocytes, in any case, continue to produce MVs throughout their life-span (Gomez et al, 1996), although not all these new MVs are TNAP-positive (Akisaka & Gay, 1985, Bonucci et al., 1992), in spite of the fact that their calcium-binding sites are invariably stained.

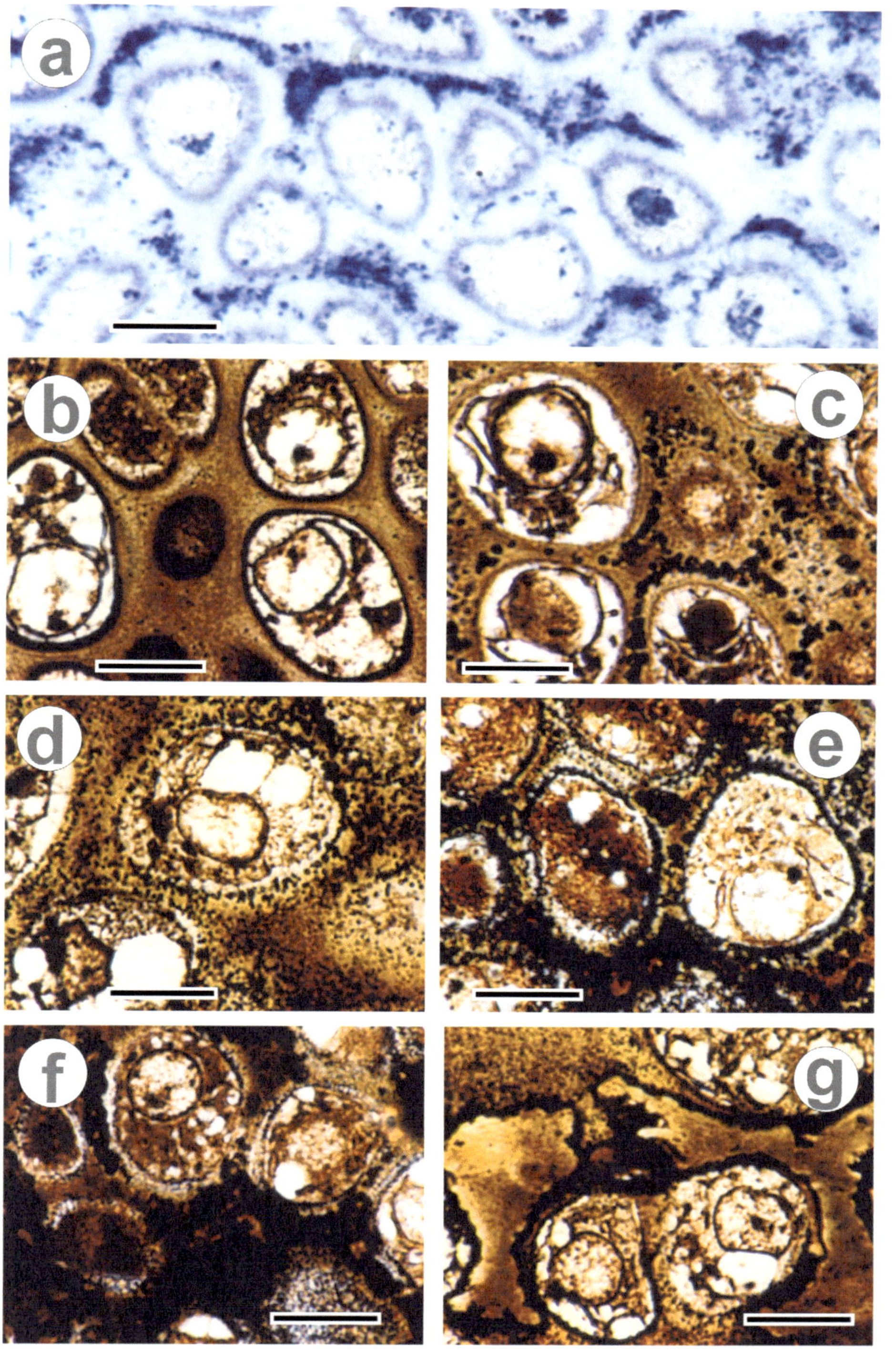

Fig. 4. TNAP (a), Calcium-binding staining (b-g) in chick embryonic cartilage. (Bar = 10 μm)

Staining for calcium-binding sites requires the incubation of slices of fresh cartilage in a solution containing 10-15mM of lanthanum chloride. This method was proposed by Morris and Appleton (1984) for electron microscopy. It was later studied in depth, at optical and electron microscopic level, using backscattered electron imaging and transmission electron microscopy (Gomez et al., 1996). The method is based on the premise that La^{3+} and Ca^{2+} have the same ionic radius but the La^{3+} has a greater charge, so that it is less easily displaced. Lanthanum allows very precise staining of the initial mineralization (Ca^{2+}-binding) sites, which then appear electron-dense.

Calcium-binding sites (after using lanthanum ions) are revealed under the light microscope by the ammoniacal silver impregnation method. In the growth plate, the upper hypertrophic chondrocytes show slightly stained dots at their peripheral membrane, and staining is also found focally in the territorial matrix (Figure 3d). In the lowest zone, the last hypertrophic chondrocytes (before apoptosis) show linear silver deposits on the peripheral membrane (Figure 3e), and the mineralizing matrix is completely stained, whereas the calcified deposits remain unstained (Figure 3f). In chick cartilage, the findings are similar in the upper territories (Figure 4 b, c), whereas in the lower zones many chondrocytes show thicker linear deposits, and a large number of dots are seen near the cells (Figure 4d-g). Matrix is stained first in the middle of the upper hypertrophic zone (Figure 4c), and is heavily stained in the lower zones (Figure 4e, f). A peripheral rim can be made out too around the bulk of the unstained calcified deposits (Figure 4g). Interestingly, the figures depicted by the La-incubation method are very similar to those displayed by confocal laser microscopy using fluorescent Ca^{2+} probes in sections of fresh chick cartilages and in cell cultures (Wu et al., 1995, 1997b). Chick chondrocytes are believed to maintain a sort of 'breathing' process by releasing calcium-Pi packets into the matrix.

The optical study has limited resolution; when electron microscopy was used, MVs proved to be stained in a rather different way. Figure 5 shows the different types of MVs stained by lanthanum and their approximate location.

MVs marked (a) are found early in the interterritorial matrix of upper zones; lanthanum staining is extravesicular and is attached to the membrane of calcified MVs. There is an ultrastructural similarity between these complexes and TNAP-rich MVs in which the reaction product is localized at the periphery. MVs marked (b) are located in the peripheral rim of the upper hypertrophic chondrocytes. They appear as globules homogenously filled with lanthanum. MVs of the last type (c) are found in large numbers around the lower chondrocytes in connection with their peripheral membranes. They are filled with lanthanum and also show numerous intravesicular densities about 10nm thick. The chondrocytes in this zone often show intracellular La-deposits.

Backscattered electron imaging of these lanthanum sites makes it possible to obtain a map of selected areas by energy dispersive X-ray analysis (Figure 6). The early lanthanum sites found in the matrix showed Sulphur (S), Phosphorus (P) and Lanthanum (La) co-localization (Figure 6a).At the pericellular rim of upper chondrocytes, mapping shows La and P co-localization (Figure 6b), whereas the lower ones show La, P and, surprisingly, Calcium (Ca) peaks (Figure 6c). All the MVs produced by hypertrophic chondrocytes bind lanthanum, and therefore have calcium-binding capabilities; in addition, the lower ones already contain calcium when released, possibly as preformed labile calcium mineral nuclei, as suggested by Wuthier and Lipscomb (2011).

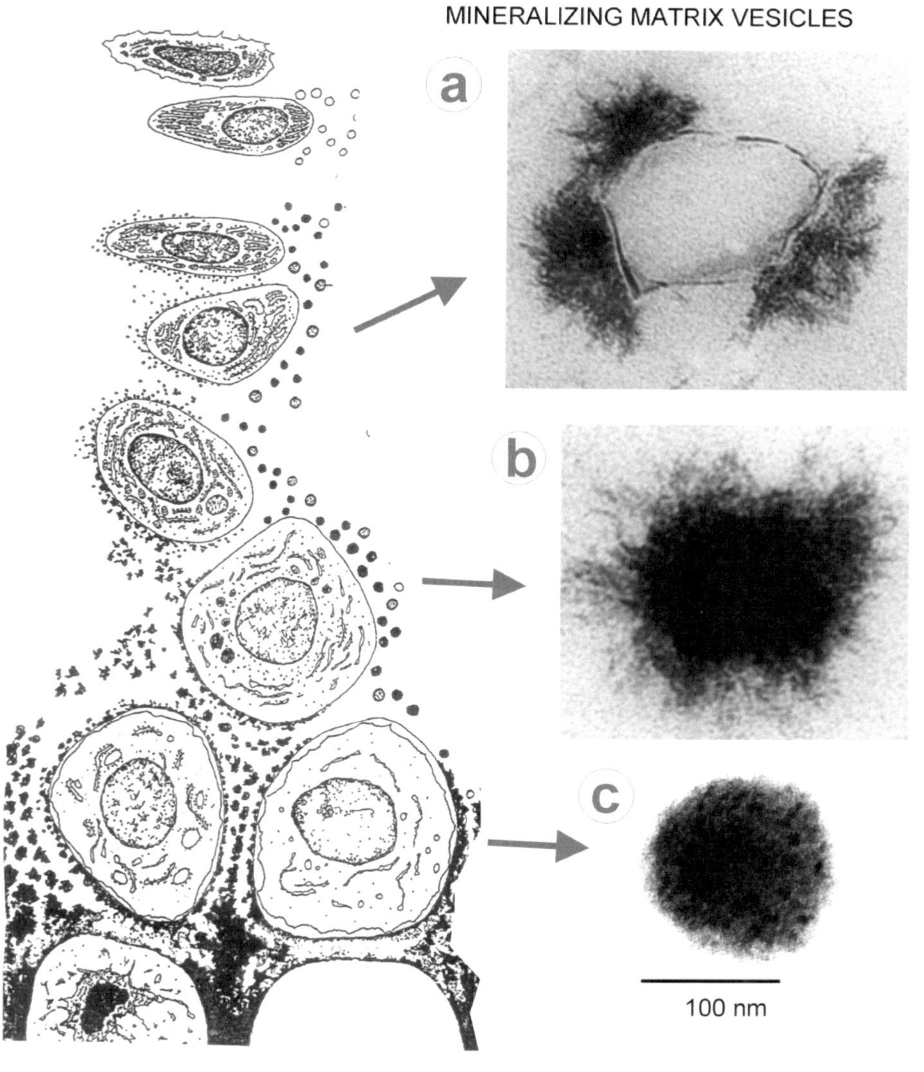

Fig. 5. Scheme of mineralizing MVs as detected by calcium-binding staining using lanthanum ions.

These morphological types of MVs, which are TNAP-rich and calcium-binding, seem to correspond to those studied by isolation methods. In fact, MVs have been separated into different density fractions corresponding to slow or quick mineralization (Warner et al., 1983). Slowly mineralizing MVs require organic phosphate substrates, and mineralization is blocked by the release of alkaline phosphatase. The activity of quickly mineralizing MVs depends on the presence of Annexin V, Ca^{2+}- protein phospholipids complexed to form an unstable mineral nucleational complex (revised by Wuthier and Lipscomb, 2011) - so they do not require organic phosphate to accumulate calcium and phosphate in vitro, and the removal of TNAP has only a minor effect.

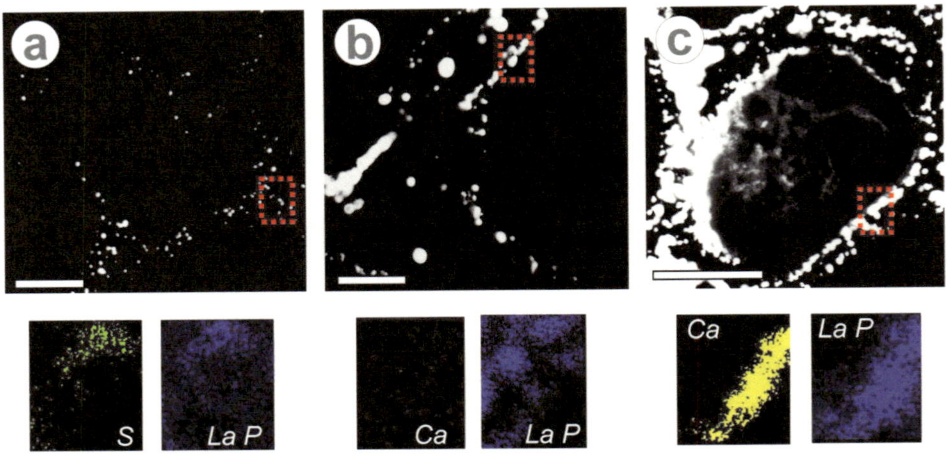

Fig. 6. Backscattered electron imaging in conjunction with elemental X-ray mapping of selected areas (red boxes) from upper matrix in chick cartilage (a); peripheral rim of hypertrophic rat chondrocytes (b); and peripheral rim of hypertrophic chick chondrocytes (c). (Bar = 5 μm)

The production of MVs during the mineralizing sequence is probably the hallmark of the mineralization process, but its mechanism is still a topic for discussion. Current concepts on MVs support two conflicting viewpoints based on mammalian and avian growth cartilage studies. *In vitro* studies of MVs isolated from the normal and rachitic rat cartilage or from normal calves have led to the conclusion that alkaline phosphatase, Ca^{2+} -ATPase, is needed for calcification (Hsu and Anderson, 1984, 1995 a, 1995b, 1996; Hsu et al., 1999; Kanabe et al., 1983). By contrast, studies on MVs isolated from chick cartilage have stressed the role of an intravesicular nucleational core (Genge et al., 1988, 1989, 1990, 1991; Kirsch et al., 1997; McLean et al., 1987; Nie et al, 1995; Register et al., 1984, 1986; Sauer & Wuthier, 1988; Wu et al., 1993, 1997a; Wuthier, 1992). Considering that chondrocytes produce various different types of MVs during the sequence, this contradiction may only be apparent, because some mechanisms could be redundant. On the other hand, re-examination of the ultrastructural micrographs accompanying Anderson's studies shows that apatite deposition is found mainly around MVs (though some MVs are mineralized within), so that the doubt arises that the TNAP-mineralizing mechanism proposed by Anderson's group should actually refer exclusively to the early mineralization of the matrix.

3. Nature of calcified deposits

The calcified deposits in cartilage are believed to be similar to those in bone, dentin and other calcified tissues: they are considered to consist of calcium phosphate and to correspond to very small crystals which, when viewed under the electron microscope, have a needle- or filament-like shape and measure from about 2 to 5 nm in thickness and from 40 to 160 nm in length. Given that structures with similar characteristic give in bone diffractograms of hydroxyapatite type, the cartilage structures, too, are usually labelled 'hydroxyapatite crystallites'. The force of habit is so strong that this denomination has been retained by many, although it is seriously misleading. First, as discussed in several reviews (Bonucci, 2007; Boskey, 1998; Veis, 2003), the inorganic deposits in cartilage - in parallel with those in bone, dentin and other hard tissues - contain, besides calcium and phosphate, other ions too, notably carbonate, magnesium, sodium, potassium and zinc. Early mineral deposits are, in fact, complexed with Zinc ions that can be found either as components of the mineral or as elements bound to metallo-enzymes such as alkaline phosphatase (Gomez et al., 1999). Second, the Ca/P molar ratio of the mineral substance is not only lower than that of hydroxyapatite, but is also variable and increases with the age of the deposits. Third, the crystalline organization is questionable, because, in our experience, and in agreement with results previously reported in bone (Landis & Glimcher, 1978), the early deposits of inorganic substance formed in epiphyseal cartilage fail to generate any electron diffraction patterns of the specific calcium phosphate solid phase, which are, in fact, only produced by the more calcified regions, whose reflections, in any case, remain those of poorly crystalline hydroxyapatite. According to Wheeler and Lewis (1977) and Arnold et al. (2001), the structures that form the early calcified deposits in bone are apatitic, but their crystal lattice contains so many distortions that they come to be intermediate between amorphous and crystalline; i.e., they have a paracrystalline character comparable with biopolymers. Fourth, the question is complicated by the possibility that amorphous calcium phosphate precedes the formation of the crystalline phase (Nudelman et al, 2010). This question is further discussed below (Chapter 5.2).

4. The matrix component related to calcification

There can be little doubt that components of the matrix are crucial for the induction and regulation of the calcification process. This is borne out by the findings that *in vitro* some of them induce the formation of hydroxyapatite, that a close association exists *in vivo* between the mineral substance and most of them (so that their total extraction can only take place after decalcification), that some are changed by the calcification process, and that their spontaneous or induced changes may cause abnormal calcification of the matrix. The specific role of each of them in calcification is, however, hard to determine, mainly because of their heterogeneity, their reciprocal interactions and their possible post-translational changes. In addition, serum proteins which permeate the matrix may have an inhibitory role during the earliest phase of the calcification process (Heiss et al, 2003).

4.1 Collagen

The results of investigations on the calcification of bone, dentin, tendons and other collagen-rich tissues have led to the conclusion that the collagen fibrils of the matrix play a leading role in the deposition of inorganic substance. In areas of initial calcification, in fact, this substance shows a close relationship with the periodic banding of collagen, due to its location within the 'holes' zone that results from the rearrangement of collagen molecules into fibrils. The development of inorganic bands exactly corresponding to the period of the collagen fibrils or, more exactly, to their holes, has led to the conclusion that calcification occurs through a process of heterogeneous nucleation catalysed by a particular atomic organization of these fibril 'holes' (Glimcher & Krane, 1968).

This theory, on which several reviews can be consulted (Bonucci, 1992, 2007; Höhling et al., 1995; Veis, 2003), can hardly be operative in the case of cartilage calcification. The typical pattern of electron-dense bands that coincide with the collagen periodic binding found in the areas of early calcification in bone has never been found in cartilage, where the early inorganic aggregates correspond to roundish calcification nodules. This agrees with the fact that cartilage contains type II collagen, whose thin fibrils, formed by homotrimers of α1(II) chains, have a poorly recognizable period and display no identifiable 'hole' zones.

This does not rule out the possibility that cartilage collagen participates in the calcification process. Calcification nodules contain chondrocalcin (reviewed by Poole et al., 1989), a calcium-binding protein associated with calcification and later identified as the C-propeptide of type II collagen (van der Rest et al., 1986). The exact role of this protein is not known; it is of interest that it is found at the beginning of the calcification process, whereas it is absent from completely calcified nodules. Calcification might also be mediated by FACIT (fibril-associated collagen with interrupted triple helices) collagens, which are characterized by the interposition of non-triple-helical domains between two or three triple-helical domains, so acquiring the possibility of association and formation of cross-links with collagen type II and with other molecules that contribute to stabilizing the matrix (Olsen, 1989). Collagen type X appears to possess the best credentials in this respect among these numerous collagens: it is specifically expressed, in fact, by hypertrophic chondrocytes (Linsenmayer et al., 1988) and its synthesis precedes matrix calcification (Iyama et al., 1991), so that it might well play a role in the process. It is covalently cross-linked to type II collagen and both bind to matrix vesicles; it binds calcium in a dose-dependent manner. Isolated matrix vesicles deprived of the associated type II and type X collagens show a marked fall in Ca uptake, which can be restored by collagen reconstitution (Kirsch & Wuthier, 1994). Type X collagen transgenic animals undergo disruption of the matrix around hypertrophic chondrocytes, anomalous proteoglycan distribution, and abnormal vertebral development (Jacenko et al., 2001). The function of type X collagen, however, remains uncertain.

4.2 Proteoglycans

The wide spaces outlined by type II collagen fibrils contain abundant non-collagenous components, the most representative of which are acid proteoglycans. These mainly consist of aggrecan, whose molecules aggregate by binding to hyaluronic acid (hyaluronan) which, in its turn, is bound to a globular link protein, so that macromolecular aggregates are formed. The composition of the aggrecan glycosaminoglycan chains varies, although there is a prevalence of chondroitin sulphate and, to a lesser degree, of keratan sulphate. An additional factor is that the protein core itself can vary, giving rise to different members of the aggrecan family (versican, neurocan, brevican). Perlecan and syndecans are cartilage proteoglycans that contain high concentrations of heparan sulphate. Decorin and biglycan have only been found in the resting cartilage.

A family of four oligomeric extracellular matrix proteins, the first and third of which are mainly expressed in cartilage, have been described as matrilins (Deák et al., 1999). They share a structure consisting of von Willebrand factor A domains, epidermal growth factor-like domains and a coiled coil alpha-helical module; post-translation proteolytic processing may cause extensive heterogeneity of their tissue forms. Matrilins contribute to the regulation of matrix assembly by binding to collagen fibrils, to other noncollagenous proteins and to aggrecan (reviewed by Klatt et al., 2011).

Acid proteoglycans have long been associated with calcification, since the early suggestion of Sobel (1955) that a complex of chondroitin sulphate and collagen in a critical conformation constitutes the 'local factor' that is responsible for calcium deposition. This hypothesis found wide support, on the basis of the observation that plenty of acid proteoglycans are found in the cartilage matrix, that they can bind high concentrations of calcium and that this can be released locally by degradation of their molecules, so creating an environment suitable for the precipitation of hydroxyapatite. By contrast, it has been shown that acid proteoglycans in solution inhibit the precipitation of calcium and phosphate (Dziewiatkowski & Majznerski, 1985) and that the breakdown of their molecules fails to trigger precipitation (Blumenthal et al., 1979). Without further considering this controversial question (see reviews by Bonucci, 2007; Roughley, 2006; Schaefer & Schaefer, 2009; Shepard, 1992), the available results suggest that the function of acid proteoglycans chiefly depends on their being aggregates or monomers, and on their hydrodynamic size.

In this connection, a number of data show that matrix proteoglycans undergo modifications pertinent to the calcification process. This has been shown by immunohistochemistry (Hirschmann & Dziewiatkowski, 1966) and confirmed by energy dispersive X-ray elemental analysis showing that matrix sulphur levels fall with calcification (from 3.5% in the uncalcified matrix to 0.3% in the fully calcified matrix; Althoff et al., 1982; Boyde & Shapiro, 1980). Lohmander and Hjerpe (1975) also found, by centrifugation of finely ground material in acetone/bromoform density gradients followed by density gradient ultracentrifugation, that the cartilage matrix loses about half its proteoglycan content with the onset of calcification, and that the proteoglycans of the calcified matrix differ in composition and size from those of uncalcified cartilage. These results were in line with electron microscope data showing that during calcification a marked decrease occurs in the size of the granules that correspond to collapsed acid proteoglycans (Buckwalter et al., 1987; Matukas & Krikos, 1968; Takagi et al., 1983, 1984;). Loss of proteoglycans with calcification was also reported by de Bernard et al. (1977), Mitchell et al. (1982), Vittur et al. (1979). On the other hand, Barckhaus et al. (1981) did not find any significant sulphur loss in frozen, freeze-dried cartilage studied by electron microscope microprobe analysis, and Scherft and Moskalewski (1984), on the basis of the number of matrix granules and the affinity of the matrix for colloidal thorium dioxide, concluded that degradation of proteoglycans is not a first, indispensable step in cartilage mineralization.

These controversial results can be settled on the basis of the autoradiographic observation, already made by Campo and Dziewiatkowski in 1963, that, although the protein of the proteoglycans is somehow removed before calcification of the cartilage, a portion of the chondroitin sulphate is retained and becomes part of the calcified matrix. Rather than being lost, the cartilage proteoglycans could be depolymerized and modified, to be finally entombed in the calcified matrix (Campo & Romano, 1986). These topics concur, and will be further considered, with those concerning crystal ghosts (Chapter 5.2).

The degradation of proteoglycan molecules might be due to the effects of enzymes (Boskey, 1992). Proteoglycan-degrading enzymes are produced by cartilage cells and their concentration is higher in the lower hypertrophic and calcification zones than in the other cartilage zones (Ehrlich et al., 1985). Most of these enzymes, such as acid phosphatase and aryl sulphatase, have a lysosomal origin (Meikle, 1975; Thyberg et al, 1975); only the former would be found in the extracellular matrix of the cartilage (Meikle, 1976). Matrix vesicles are also selectively enriched in enzymes which degrade proteoglycans (Dean et al., 1992), although their most typical enzyme is alkaline phosphatase.

4.3 Alkaline phosphatase (TNAP)

Since the earliest suggestion of Robison (1923), that the formation of calcium phosphate might be dependent on the hydrolysis of phosphate esters by alkaline phosphatase, plenty of studies have been centred on this enzyme (reviewed by Orimo, 2010; Wuthier and Lipscomb, 2011) whose importance has been heightened by its already discussed relationship with MVs (Chapter 2). Its function in the calcification process, however, remains controversial, although it is certainly fundamental in all processes of biological calcification. This topic is further discussed below (Chapter 5.1).

4.4 Glycoproteins and phospholipids

The cartilage matrix positively reacts when treated with the periodic acid-Schiff method, a histochemical method that shows molecules having vicinal glycol groups, i.e., glycoproteins. Some of these molecules are phosphorylated and expressed by both chondrocytes and osteoblasts, as in the case of osteopontin and bone sialoprotein. The localization of these glycoproteins, and their function in calcification, remain rather elusive (reviewed by Gentili & Cancedda, 2009).

Phospholipids too are components of the cartilage matrix and are probably involved in the calcification process, as shown by the accumulation of ^{32}P-orthophosphate at the calcification front and by the observation that a fraction of them can only be extracted after decalcification (Eisenberg et al., 1970). Their incorporation in the calcified matrix has been confirmed by immunohistochemistry using MC22-33F, an antibody that recognizes phosphatidylcholine, sphingomyelin and dimethylphosphatidylethanolamine, and that gives a strong reaction at the periphery of the calcification nodules (Bonucci et al., 1997). Further confirmation was obtained by combining malachite green fixation with the complex phospholipase A_2-gold; again the reaction was stronger at the periphery than at the centre of the calcification nodules (Silvestrini et al., 1996). These results, together with the long-standing knowledge that bone and other calcifying tissues contain calcium-phospholipid-phosphate complexes (Boskey & Posner, 1976) and that in calcifying cartilage the interaction of Ca and P ions with phosphatidylserine can give rise to phospholipid-Ca- Pi complexes (Boyan et al., 1989), make the phospholipids good candidates as molecules capable of inducing or controlling the calcification process.

5. Matrix calcification

The calcification of the cartilage has long been considered a straightforward process comprising just one single phase– the precipitation of calcium phosphate in specific areas of the matrix. In this area of research, things turn out to be much more complicated, and the process actually goes forward through at least three phases: the intervention of specific calcifying structures, the development of crystal-like, organic-inorganic particles, and their gradual transformation into inorganic structures that finally mature into hydroxyapatite crystallites.

5.1 First phase: Structures that undergo calcification

A number of electron microscope studies have shown that in epiphyseal cartilage the early mineral aggregates are connected to MVs, which are often partly or totally filled by crystals, or may be found in contact with peripheral crystal aggregates. The sheer abundance of MVs suggests that there are many crystal centres from which mineralization spreads out into the matrix. In reality, most of the early matrix mineralization begins around MVs (Figure 7), i.e., is an extravesicular process, which is probably favoured by the existence of a mineralization centre (none other than the intravesicular mineral), but is regulated by a second mineralizing mechanism. This is suggested by the observation that there are pathological conditions - hypophosphatasia, in particular - in which the mineralization process occurs within and around MVs, but fails to spread into the surrounding matrix (Anderson et al., 1997). For mineralization to spread beyond MVs, therefore, a second, facilitating mechanism is required.

The possible role of collagen fibrils, acid proteoglycans, glycoproteins, and phospholipids in calcification has been discussed above. These structures are components of the uncalcified matrix; some type of modification therefore seems necessary for them to be able to take part in the calcification process. Removal of inhibitors might be one, molecular changes of various types might also be involved. Unfortunately, these processes are poorly known (reviewed by Boskey, 1992). The role of TNAP in calcification must obviously be emphasized: as already mentioned, any serious lack of TNAP (as in hypophosphatasia) prevents the calcification process from spreading beyond MVs into the matrix. It is present in all tissues that calcify, and is located precisely in the areas that will calcify, so much so that its histochemical reaction product gives an ultrastructural picture similar to that of the mineral substance around MVs. The role of TNAP, however, is still in doubt,

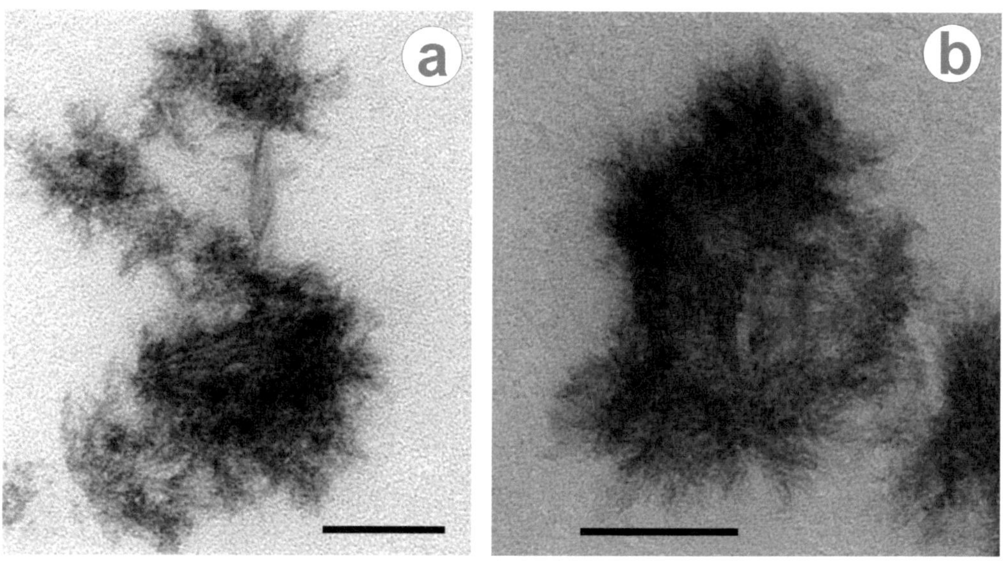

Fig. 7. Early matrix mineralization sites around calcified MVs as shown by calcium-binding staining using lanthanum ions. (Bar = 100 nm)

and as many as six different functions have been attributed to the enzyme (Wuthier & Register, 1985), the two most often put forward being an increase in the local concentration of phosphates, which would permit hydroxyapatite formation, and the removal of inorganic pyrophosphate (PPi), which is an inhibitor of calcification. Another function is often neglected: the TNAP of MVs from growth cartilage is a calcium-binding glycoprotein (de Bernard et al., 1985), a property that might permit crystal formation through the mechanism depicted below for crystal ghosts (Chapter 5.2). In this context, it is interesting that a zinc-containing glycoprotein can be demonstrated in the matrix of calcified deposits (Gomez et al, 1999). Molecular studies have shown that TNAP contains a specific metal binding domain (different from the zinc-active site). A synchrotron radiation X-ray fluorescence study confirmed that the metal in the metal-binding site is a calcium ion (Mornet et al., 2001).

5.2 Second phase: Development of organic-inorganic, crystal-like structures

The structures that develop within and around matrix vesicles during the early calcification phase are usually called 'crystallites', although, as reported above (Chapter 3), they appear to be intermediate between amorphous and crystalline, i.e., they have a paracrystalline status comparable with biopolymers. Because of their prevalent origin from matrix vesicles, most of them are collected in roundish aggregates called 'calcification nodules'. Electron microscope studies have contributed much to the knowledge of these structures. As already mentioned in Chapter 3, the cartilage 'crystallites' appear under the electron microscope as filament- and needle-like structures, which have intrinsic electron-density; as a result, they do not need to be stained to become visible on the microscope screen. This advantage, which is directly attributable to their inorganic content, is counterbalanced by the fact that their electron density completely masks the organic structures they are associated with. This masking effect, which prevents recognition of the organic components of the calcified matrix and their relationship with the inorganic substance, can only be eliminated by decalcification. This procedure, however, leads to the removal not only of the inorganic material, but also of a number of organic molecules, so that the decalcified areas appear as almost empty zones crossed by collagen fibrils (reviewed by Bonucci, 2007). This disadvantage can be overcome by using special decalcification techniques such as the PEDS method and the cationic dye stabilization method.

5.2.1 'Crystal ghosts'

The acronym PEDS stands for Post-Embedding Decalcification and Staining, a method that, unlike the usual decalcification by immersion of whole specimens in the decalcifying solution, followed by dehydration and embedding, decalcifies the tissue after its embedding in a resin, that is, by floating ultrathin sections on the surface of the decalcifying solution (Bonucci & Reurink, 1978). The extreme thinness of the sections (usually less than 1 mμ) allows their decalcification to become complete in only a few minutes. At the same time, because the tissue is embedded in the resin, its organic components are blocked and stabilized, and therefore protected from solubilization and extraction. The preservation of the organic structures is confirmed by the ultrastructure of the cells and of the uncalcified matrix: this is conspicuously altered by the usual methods of decalcification, while after the PEDS method it is indistinguishable from the ultrastructure these structures show in the undecalcified sections. Strangely enough, the calcification nodules appear electron-dense and contain filament- and needle-like structures similar to untreated 'crystallites'. This effect, which at first glance may appear to be due to the lack of decalcification and to the persistence of the so-called 'crystallites', must actually be put down to the fact that organic structures previously masked by the mineral substance have become unmasked by decalcification and have then been stained. The unexpected finding is that these structures have practically the same shape and size as the 'crystallites' and must consequently be considered as their 'organic ghosts' (Figure 8). This is why they were first called 'crystal ghosts'. As a consequence, the early 'crystallites' found in the calcification nodules must be considered organic-inorganic hybrids, each consisting of an organic filament with attached calcium and phosphate ions.

Confirmation of these findings has come from studies that have adopted the second decalcification method mentioned above, i.e., the cationic dye stabilization method (reviewed by Bonucci, 2002). This is based on the notion that acid polyanions react with basic substances like cationic dies, so that the latter can be used to reveal the former in a tissue (according to the process often referred to as 'basophilia'). The same staining substances stabilize the acid molecules, so that these become insoluble and resistant to decalcification, and do not collapse into granules during dehydration. The ultrastructural study of cartilage treated with cationic dies, after its decalcification by immersion in a decalcifying solution, and later embedding in a resin, shows that the calcification nodules appear as aggregates of organic, filament-like structures which bear a close resemblance to 'crystallites' on one hand and to crystal ghosts on the other (Figure 8a). The importance of these findings is not limited to their confirmation of the existence of crystal ghosts, but includes their demonstration that crystal ghosts cannot depend, as suggested by Dong and Warshawsky (1995), on the penetration by staining heavy metals of the voids left in the resin by the dissolution of the crystallites, because the decalcification procedure is always carried out before embedding.

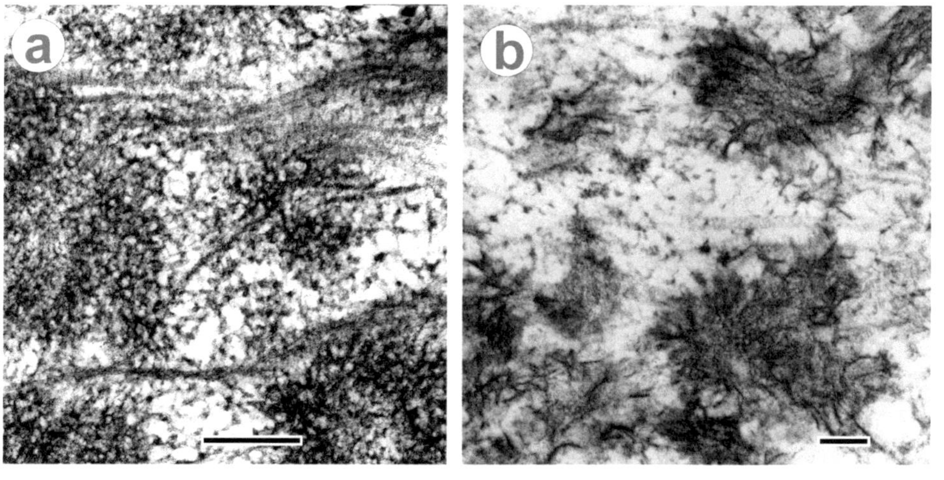

Fig. 8. (a) Crystal ghosts in rat growth plate after acridine orange dye stabilization and EDTA decalcification; (b) Crystal ghosts in embryonic chick cartilage after PEDS. (Bar = 200nm)

5.2.2 The nature of crystal ghosts

The results reported in the previous chapter show that the early 'crystallites' of the calcifying cartilage are organic-inorganic hybrids and that the crystal ghosts represent their organic component. On this basis, it becomes mandatory to establish their nature and composition.

As reported in Chapter 3.4.2, the acid proteoglycans are the most abundant components of the cartilage matrix and are responsible for its long-recognized basophilia and metachromasia. It is true that calcification slightly reduces staining properties, but they do, on the whole, persist, which shows that the calcification nodules, too, contain acid proteoglycans. Because the crystal ghosts are the most abundant components of the calcification nodules, the logical step forward is to suppose that they are proteoglycan molecules.

This hypothesis is strongly supported by electron microscope histochemistry. The crystal ghosts, in fact, react with, and are stained by, ruthenium red, ruthenium hexammine trichloride, terbium chloride, colloidal iron, or bismuth nitrate (Bonucci et al., 1989; Bonucci, 2002). All these histochemical reactions can occur at pH as low as 1.8, showing that the organic substrates have strong acidic groups, such as the sulphate groups of acid proteoglycans. The possibility that the crystal ghosts of the calcified cartilage are acid proteoglycans is confirmed by the observation that the histochemical reactions are inhibited by methylation, which blocks sulphate and carboxy groups, and are not restored by saponification, which only re-establishes the carboxy groups (Bonucci et al., 1988). Further confirmation is given by the immunoreaction of crystal ghosts with the antibody CS-56, which is specific for the glycosaminoglycan portion of chondroitin sulphate (Bonucci & Silvestrini, 1992). It must be added that crystal ghosts also react with acidic phosphotungstic acid, which is a glycoprotein stain.

5.2.3 The function of crystal ghosts

The relationship between crystal ghosts and inorganic substance in the 'crystallites' is so close that they cannot be distinguished from one another under the electron microscope, even if the former are stained by heavy metals (uranyl acetate, lead citrate). The hybrids that they form appear as unique structures; only after decalcification can their organic component be made out. It is obvious that this close organic-inorganic relationship necessarily brings with it strong implications.

Actually, as reported in Chapter 4.2, acid proteoglycans have long been considered to be responsible for the calcification process in cartilage. The identification of crystal ghosts as molecules of acid proteoglycans strongly supports this possibility. The acid groups of these molecules can bind high concentrations of calcium and give rise, with the possible initial formation of amorphous calcium phosphate, to the organic-inorganic structures that are commonly called 'crystallites'. The proteoglycan molecules could function as templates, and the 'crystallite' shape and size could simply reflect the filamentous shape and size of their organic framework. In this connection, the suggestions must be taken into consideration that prenucleation clusters of calcium carbonate may be stabilized by organic molecules (Gebauer et al, 2008) and that the template-directed aggregation of these clusters give rise to the formation of amorphous calcium carbonate nanoparticles which assemble at the template and develop into crystalline domains (Pouget et al, 2009).

Crystal ghosts have been described not only in the calcifying cartilage, but in similar terms in other hard tissues during the early stage of calcification (reviewed by Bonucci, 2007). The acid proteoglycans are ubiquitous in these tissues and it might be thought that they behave as crystal ghosts in all of them. It must be noted, though, that other acidic molecules could play exactly the same role; a number of polyanions, some characterized by the repetitive sequences of aspartic acid, have been found in all calcified tissues and each of them could theoretically initiate and regulate the calcification process (Gotliv et al., 2003; Rahman & Oomori, 2010; Takeuchi et al., 2005; Weiner & Addadi, 1991). Moreover, it cannot be disregarded that, as reported above (Chapter 3.5.1), TNAP, too, is a glycoprotein with calcium-binding properties. It can be demonstrated histochemically in all areas of the matrix undergoing calcification, whereas it is left unstained, even if still present, in already calcified areas. It may be speculated, therefore, that TNAP can bind mineral ions, as crystal ghosts do, so becoming deactivated and embedded in the calcified matrix.

5.2.4 Are there pre-crystal ghosts?

If crystal ghosts have the role suggested above, then the question arises whether they are preformed in the matrix and exactly what kind of mechanism induces their activation as mineral-binding structures. It may be hypothesized that the matrix proteoglycans, and the other proteins mentioned above, must in some way be modified to acquire calcium-binding properties; the great variety of proteolytic enzymes located in the cartilage might well carry out this function. The reaction of lanthanum with components of the matrix seems to strengthen this hypothesis (Figure 9).

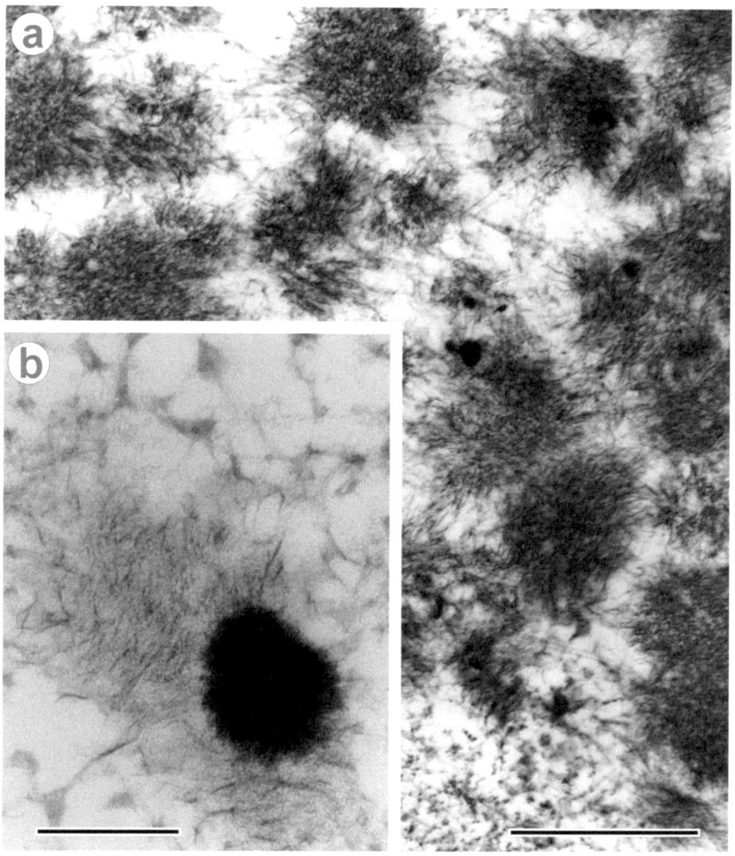

Fig. 9. (a) Pre-crystal ghosts in rat growth plate (bar = 1 μm); (b) and in chick embryonic cartilage, (bar = 200 nm)

Staining the growth cartilage with lanthanum chloride unexpectedly shows that the still uncalcified matrix of the upper chondrocyte zones contains roundish aggregates of very fine filaments which closely resemble crystal ghosts and, on the whole are similar to calcification nodules (Gomez et al., 1996). They correspond to focal concentrations of proteoglycans with a high La-binding capacity, or, in a broader perspective, strong calcium-binding properties. These findings are in agreement with the hypothesis that the acid proteoglycans of the cartilage matrix, which are inhibitors in their native state, can become inducers of the calcification process after their molecules have in some way become modified. The lanthanum-stained structures might be pre-crystal ghosts, that is, organic molecules that are ready to bind calcium ions and so trigger 'crystallite' formation.

5.3 Third phase: Changes in crystallites as mineralization progresses

The progression of the calcification process implies that the early calcification nodules acquire new 'crystallites' at their periphery, so increasing in size and gradually coalescing with each other till the whole matrix is calcified. As a consequence, the developing calcification nodules become fully calcified in their central area while they are still calcifying at their periphery, where 'crystallites' continue to be formed. Another process occurs at the same time: the PEDS method shows that, as the calcification nodules enlarge, the crystal ghosts disappear from their central, fully calcified zones, where all that can be recognized is an amorphous material; the crystal ghosts remain visible at the periphery of the nodules, where the process of calcification is still under way (Figure 8b). The completion of the calcification process takes place, therefore, in parallel with a loss of the organic components that constitute the crystal ghosts. In agreement with these observations, the loss of organic material during calcification has been found biochemically not only in cartilage (Lohmander and Hjerpe, 1975; Vittur et al., 1979), but in other calcified tissues, too, especially in bone (Pugliarello et al., 1970) and enamel (reviewed by Bartlett & Simmer, 1999; Simmer& Hu, 2002). These results suggest that, as calcification progresses, the organic components of the 'crystallites', recognizable as crystal ghosts, are gradually lost, probably through the lytic effects of proteases. The fall in amounts of organic material as the calcification process is completed can be seen as converging with other two processes mentioned in the preceding chapters: first, the Ca/P molar ratio increases and approaches that of hydroxyapatite; second, the electron diffractograms, which are of amorphous type if obtained from the early, small, incompletely developed calcification nodules, become of poorly crystalline type if obtained from the central zone of the biggest nodules or from the already diffusely calcified matrix. The loss of crystal ghosts therefore seems to be a pre-requisite for the inorganic component of 'crystallites' to acquire a definitive hydroxyapatite organization.

6. Conclusion

The results reported above allow a few conclusions to be drawn. The calcification of the cartilage matrix seems to occur through two mechanisms, one related to the development of matrix vesicles, another involving TNAP and other matrix components. The first is characterized by the formation of early mineral aggregates within MVs, mostly through their annexin calcium-binding properties; the second consists in the formation of organic-inorganic hybrids and implies the activation of alkaline phosphatase. These hybrids, incorrectly called 'crystallites', undergo a process of transformation which includes the gradual loss of their organic component, the increase of their Ca/P molar ratio, and the transformation of their electron diffractograms from amorphous to crystalline. It seems that the acidic molecules that function as templates allow the linkage and organization of inorganic ions along planes that approximate those of hydroxyapatite, and that their enzymatic breakdown permits the ions to move all over the definitive hydroxyapatite reticulum. The organic components of the 'crystallites' are pre-formed in the matrix and must in some way be activated, probably by metalloproteases derived from lysosomes and MVs. These changes are most directly pertinent to acid proteoglycans which, as shown by the lanthanum reaction, acquire a focal capacity to bind calcium and phosphate ions.

Although most of these concepts must still be regarded as speculative, they appear to offer a rational explanation for all the main aspects of the calcification mechanism in cartilage - an explanation whose basic features are probably applicable to all calcifying tissues.

7. References

Akisaka, T. & Gay, CV. (1985). Ultrastructural localization of calcium-activated adenosine triphosphatase (Ca2+-ATPase) in growth-plate cartilage. *J Histochem Cytochem* 33:925-932.

Allerstorfer, D.; Longato, S.; Schwarzer, C.; Fischer-Colbrie, R.; Hayman, A.R. & Blumer, M.J. (2010). VEGF and its role in the early development of the long bone epiphysis. *J Anat* 216:611-624

Althoff, J.; Quint, P.; Krefting, E.-R. & Höhling, H.J. (1982). Morphological studies on the epiphyseal growth plate combined with biochemical and X-ray microprobe analyses. *Histochemistry* 74:541-552

Anderson, H.C. (1967). Electron microscopic studies of induced cartilage development and calcification. *J Cell Biol* 35:81-101

Anderson, H.C. (1969). Vesicles associated with calcification in the matrix of epiphyseal cartilage. *J Cell Biol* 41:59-72

Anderson, H.C. (1995). Molecular biology of matrix vesicles. *Clinical Orthopaedics* 314: 266-280

Anderson, H.C.; Hsu, H.H.; Morris, D.C.; Fedde, K.N. & Whyte, M.P. (1997). Matrix vesicles in osteomalacic hypophosphatasia bone contain apatite-like mineral crystals. *Am J Pathol* 151: 1555-1561

Arnold, S.; Plate, U.; Wiesmann, H.-P.; Straatmann, U.; Kohl, H. & Höhling, H.J. (2001). Quantitative analyses of the biomineralization of different hard tissues. *J Microsc* 202: 488-494

Barckhaus, R.H.; Krefting, E.-R.; Althoff, J.; Quint, P. & Höhling, H.J. (1981). Electron-microscopic microprobe analysis on the initial stages of mineral formation in the epiphyseal growth plate. *Cell Tissue Res* 217: 661-666

Bartlett, J.D. & Simmer, J.P. (1999). Proteinases in developing dental enamel. *Crit Rev Oral Biol Med* 10: 425-441

Blumenthal, N.C.; Posner, A.S.; Silverman, L.D. & Rosenberg, L.C. (1979). Effect of proteoglycans on in vitro hydroxyapatite formation. *Calcif Tissue Int* 27: 75-82

Bonucci, E. (1967). Fine structure of early cartilage calcification. *J Ultrastruct Res* 20:33-50

Bonucci, E. (1970). Fine structure and histochemistry of "calcifying globules" in epiphyseal cartilage. *Z Zellforsch Mikrosk Anat* 103:192-217

Bonucci, E. (1992). Role of collagen fibrils in calcification, In: *Calcification in biological systems*, Bonucci, E. (ed.), pp. 19-39, CRC Press, Boca Raton

Bonucci, E. (2002). Crystal ghosts and biological mineralization: fancy spectres in an old castle, or neglected structures worthy of belief? *J Bone Miner Metab* 20:249-265

Bonucci, E. (2007). Biological calcification. Normal and pathological processes in the early stages. Springer, Berlin, Heidelberg

Bonucci, E. & Reurink, J. (1978). The fine structure of decalcified cartilage and bone: A comparison between decalcification procedures performed before and after embedding. Calcif Tissue Res 25:179-190

Bonucci, E. & Silvestrini, G. (1992). Immunohistochemical investigation on the presence of chondroitin sulfate in calcification nodules of epiphyseal cartilage. *Eur J Histochem* 36:407-422

Bonucci, E.; Silvestrini, G. & Bianco, P. (1992) Extracellular alkaline phosphatase activity in mineralizing matrices of cartilage and bone: ultrastructural localization using a cerium-based method. *Histochemistry* 97:323-327

Bonucci, E.; Silvestrini, G. & Di Grezia, R. (1988). The ultrastructure of the organic phase associated with the inorganic substance in calcified tissues. *Clin Orthop* 233: 243-261

Bonucci, E.; Silvestrini, G. & Di Grezia,R. (1989). Histochemical properties of the "crystal ghosts" of calcifying epiphyseal cartilage. *Connect Tissue Res* 22:43-50

Bonucci, E.; Silvestrini, G. & Mocetti, P. (1997). MC22-33F monoclonal antibody shows unmasked polar head groups of choline-containing phospholipids in cartilage and bone. *Eur J Histochem* 41:177-190

Boskey, A.L. (1992). Mineral-matrix interactions in bone and cartilage. *Clin Orthop*. 281: 244-274

Boskey, A.L. (1998). Biomineralization: conflicts, challenges, and opportunities. *J Cell Biochem* 30/31 (suppl.): 83-91, 1998

Boskey, A.L.; Doty, S.B. & Binderman, I. (1994). Adenosine 5′-triphosphate promotes mineralization in differentiating chick limb-bud mesenchymal cell cultures. *Microsc Res Tech* 28:492-504

Boskey, A.L. & Posner, A.S. (1976). Extraction of calcium-phospholipid-phosphate complex from bone. *Calcif Tissue Res* 19: 273-283

Boyan, B.D.; Schwartz, Z.; Swain, L.D. & Khare, A. (1989). Role of lipids in calcification of cartilage. *Anat Record* 224: 211-219

Boyde, A. & Shapiro, I.M. (1980). Energy dispersive X-ray elemental analysis of isolated epiphyseal growth plate chondrocyte fragments. *Histochemistry* 69: 85-94

Buckwalter, J.A.; Rosenberg, L.C. & Ungar, R. (1987). Changes in proteoglycan aggregates during cartilage mineralization. *Calcif Tissue Int* 41: 228-236

Campo, R.D. & Dziewiatkowski, D.D. (1963). Turnover of the organic matrix of cartilage and bone as visualized by autoradiography. *J Cell Biol* 18:19-29

Campo, R.D. & Romano, J.E. (1986). Changes in cartilage proteoglycans associated with calcification. *Calcif Tissue Int* 39:175-184

de Bernard, B.; Bianco, P.; Bonucci, E.; Costantini, M.; Lunazzi, G.C.; Martinuzzi, P.; Modricky, C.; Moro, L.; Panfili, E.; Pollesello, P.; Stagni, N. & Vittur, F. (1986). Biochemical and immunohistochemical evidence that in cartilage an alkaline phosphatase is a Ca^{2+} -binding glycoprotein. *J Cell Biol* 103:1615-1623 de Bernard, B.; Stagni, N.; Colautti, I.; Vittur, F. & Bonucci, E. (1977). Glycosaminoglycans and endochondral calcification. *Clin Orthop* 126:285-291

Deák, F.; Wagener, R.; Kiss, I. & Paulsson, M. (1999). The matrilins: a novel family of oligomeric extracellular matrix proteins. *Matrix Biol* 18:55-64

Dean, D.D.; Schwartz, Z.; Muniz, O.E.; Gomez, R.; Swain, L.D.; Howell, D.S. & Boyan, B.D. (1992). Matrix vesicles are enriched in metalloproteinases that degrade proteoglycans. *Calcif Tissue Int* 50:342-349

Dong, W. & Warshawsky, H. (1995). Failure to demonstrate a protein coat on enamel crystallites by morphological means. *Archs Oral Biol* 40:321-330

Dziewiatkowski, D.D. & Majznerski, L.L. (1985). Role of proteoglycans in endochondral ossification: inhibition of calcification. *Calcif Tissue Int*. 37:560-564

Ehrlich, M.G.; Tebor, G.B.; Armstrong, A.L. & Mankin, H.J. (1985). Comparative study of neutral proteoglycanase activity by growth plate zone. *J Orthop Res* 3:269-276

Eisenberg, E.; Wuthier, R.E.; Frank, R.B. & Irving, J.T. (1970). Time study of *in vivo* incorporation of 32P orthophosphate into phospholipids of chicken epiphyseal tissues. *Calcif Tissue Res* 6:32-48

Farnum, C.E.; Turgai,J. & Wilsman, N.J. (1990). Visualization of living terminal hypertrophic chondrocytes of growth plate cartilage *in situ* by differential interference contrast microscopy and time-lapse cinematography. *J Orthop Res* 8:750-763

Gebauer, D., Völkel, A. & Cölfen, H. (2008). Stable prenucleation calcium carbonate clusters. *Science* 322: 1819-1822.

Genge, B.R.; Sauer, G.R.; Wu, L.N.Y.; McLean, F.M. & Wuthier, R.E. (1988). Correlation between loss of alkaline phosphatase activity and accumulation of calcium during matrix vesicle-mediated mineralization. *J Biol Chem* 263:18513-18519

Genge, B.R.; Wu, L.N.Y. & Wuthier, R.E. (1989). Identification of phospholipid-dependent calcium-binding proteins as constituents of matrix vesicles. *J Biol Chem* 264:10917-10921

Genge, B.R.; Wu, L.N.Y. & Wuthier, R.E. (1990). Differential fractionation of matrix vesicles proteins. Further characterization of the acidic phospholipid-dependent Ca^{2+}-binding proteins. *J Biol Chem* 265:4703-4710

Genge, B.R.; Wu, L.N.Y.; Adkisson, H.D. & Wuthier, R.E. (1991). Matrix vesicles annexins exhibit proteolipid-like properties. Selective partitioning into lipophilic solvents under acidic conditions. *J Biol Chem* 266:10678-10685

Gentili, C. & Cancedda, R. (2009). Cartilage and bone extracellular matrix. *Curr Pharm Des* 15:1334-1348

Glimcher, M.J. & Krane, S.M. (1968). The organization and structure of bone, and the mechanism of calcification, In: *Biology of collagen*, Gould, B.S. (ed.), pp. 67-251, Academic Press, London,

Gomez, S.; Lopez-Cepero, J.M.; Silvestrini, G.; Mocetti, P. & Bonucci, E. (1996). Matrix vesicles and focal proteoglycan aggregates are the nucleation sites revealed by the lanthanum incubation method: a correlated study on the hypertrophic zone of the rat epiphyseal cartilage. *Calcif Tissue Int* 58:273-282

Gomez, S.; Rizzo, R.; Pozzi-Mucelli, M.; Bonucci, E. & Vittur, F. (1999). Zinc mapping in bone tissues by histochemistry and synchrotron radiation-induced X-ray emission: Correlation with the distribution of alkaline phosphatase. *Bone* 25:33-38

Gotliv, B.A.; Addadi, L. & Weiner, S. (2003). Mollusk shell acidic proteins: in search of individual functions. *Chembiochem* 4:522-529

Hall, B.K. (2005). Dedifferentiation provides progenitor cells for jaws and long bones, In: *Bones and cartilage: Developmental and evolutionary skeletal biology*, Hall, B.K. (ed.), pp 166-182, Elsevier Academic Press, Amsterdam

Heiss, A., DuChesne, A., Dencke, B., Grötzinger, J., Yamamoto, K., Renné, T. & Jahnen-Dechent, W. (2003). Structural basis of calcification inhibition by α_2-HS glycoprotein/fetuin A. *J Biol Chem* 278, 5: 13333-13341.

Hirschman, A. & Dziewiatkowski, D.D. (1966). Protein-polysaccharide loss during endochondral ossification: immunochemical evidence. *Science* 154: 393-395

Höhling, H.J.; Arnold, S.; Barckhaus, R.H.; Plate, U. & Wiesmann, H.P. (1995). Structural relationship between the primary crystal formation and the matrix macromolecules in different hard tissues. Discussion of a general principle. *Connect Tissue Int.* 33:171-178

Hsu, H.H. & Anderson, H.C. (1984). The deposition of calcium pyrophosphate and phosphate by matrix vesicles isolated from fetal bovine epiphyseal cartilage. *Calcif Tissue Int* 36:615-621

Hsu, H.H. & Anderson, H.C. (1995a). Effects of Zinc and divalent cation chelators on ATP hydrolysis and Ca deposition by rachitic rat matrix vesicles. *Bone* 17:473-477

Hsu, H.H. & Anderson, H.C. (1995b). A role for ATPase in the mechanisms of ATP-dependent Ca and phosphate deposition by isolated rachitic matrix vesicles. *Int J Biochem Cell Biol* 27:1349-1356

Hsu, H.H. & Anderson, H.C. (1996). Evidence of the presence of a specific ATPase responsible for ATP-initiated calcification by matrix vesicles isolated from cartilage and bone. *J Biol Chem* 271:26383-26388

Hsu, H.H.; Camacho, N.P. & Anderson, H.C. (1999). Further characterization of ATP-initiated calcification by matrix vesicles isolated from rachitic rat cartilage. Membrane perturbation by detergents and deposition of calcium pyrophosphate by rachitic matrix vesicles. *Biochim Biophys Acta* 1416:320-332

Hunziker, E.B.; Herrmann, W.; Schenk, R.K.; Mueller, M. & Moor, H. (1984). Cartilage ultrastructure after high pressure freezing, freeze substitution, and low temperature embedding. I. Chondrocyte ultrastructure--implications for the theories of mineralization and vascular invasion. *J Cell Biol* 98:267-276

Iyama, K.; Ninomiya, Y.; Olsen, B.R.; Linsenmayer, T.F.; Trelstad, R.L. & Hayashi, M. (1991). Spatiotemporal pattern of type X collagen gene expression and collagen deposition in embryonic chick vertebrae undergoing endochondral ossification. *Anat Record* 229:462-472

Jacenko, O.; Chan, D.; Franklin, A.; Ito, S.; Underhill, C.B.; Bateman, J.F. & Campbell, M.R. (2001). A dominant interference collagen X mutation disrupts hypertrophic chondrocyte pericellular matrix and glycosaminoglycan and proteoglycan distribution in transgenic mice. *J Pathol* 159:2257-2269

Kanabe, S.; Hsu, H.H.; Cecil, R.N. & Anderson, H.C. (1983). Electron microscopic localization of adenosine triphosphate (ATP)-hydrolyzing activity in isolated matrix vesicles and reconstituted vesicles from calf cartilage. *J Histochem Cytochem* 31:462-470

Kirsch, T.; Nah, H-D.; Shapiro, I.M. & Pacifici, M. (1997). Regulated production of mineralization-competent matrix vesicles in hypertrophic chondrocytes. *J Cell Biol* 137:1149-1160.

Kirsch, T. & Wuthier, R.E. (1994). Stimulation of calcification of growth plate cartilage matrix vesicles by binding to type II and X collagens. *J Biol Chem* 269:11462-11469.

Klatt, A.R.; Becker, A.K.; Neacsu, C.D.; Paulsson, M. & Wagener, R. (2011). The matrilins: modulators of extracellular matrix assembly. *Int J Biochem Cell Biol* 43:320-330

Landis, W.J. & Glimcher, M.J. (1978). Electron diffraction and electron probe microanalysis of the mineral phase of bone tissue prepared by anhydrous techniques. *J Ultrastruct Res* 63:188-223

Linsenmayer, T.F.; Eavey, R.D. & Schmid, T.M. (1988). Type X collagen: a hypertrophic cartilage-specific molecule. *Pathol Immunopathol Res* 7:14-19

Lohmander, S. & Hjerpe, A. (1975). Proteoglycans of mineralizing rib and epiphyseal cartilage. *Biochim Biophys Acta* 404:93-109

Matsuzawa, T. & Anderson, H.C. (1971). Phosphatases of epiphyseal cartilage studied by electron microscopic cytochemical methods. *J Histochem Cytochem* 19:801-808

Matukas, J. & Krikos, G.A. (1968). Evidence for changes in protein polysaccharide associated with the onset of calcification in cartilage. *J Cell Biol* 39:43-48

McLean, F.M.; Keller, P.J.; Genge, B.R.; Walters, S.A. & Wuthier, R.E. (1987). Disposition of preformed mineral in matrix vesicles. Internal localization and association with alkaline phosphatase. *J Biol Chem* 262:10481-10488

Meikle, M.C. (1975). The distribution and function of lysosomes in condylar cartilage. *J Anat* 119: 85-96

Meikle, M.C. (1976). The mineralization of condylar cartilage in the rat mandible: an electron microscopic enzyme histochemical study. *Arch Oral Biol* 21:33-43

Mitchell, N.; Shepard, N. & Harrod, J. (1982). The measurement of proteoglycan in the mineralizing region of the rat growth plate. An electron microscopic and X-ray microanalytical study. *J Bone Joint Surg* 64A: 32-38

Mornet, E.; Stura, E.; Lia-Balidini, A.-S.; Stigbrand, T.; Ménez, A. & Le Du, M.-H. (2001). Structural evidence for a functional role of human tissue nonspecific alkaline phosphatase in bone mineralization. *J Biol Chem* 276: 31171-31178

Morris, D.C. & Appleton, J. (1984). The effects of lanthanum on the ultrastructure of hypertrophic chondrocytes and the localization of lanthanum precipitates in condylar cartilages of rats fed on normal and rachitogenic diets. *J Histochem Cytochem* 32:239-247

Nie, D.; Genge, B.R.; Wu, L.N.Y. & Wuthier, R.E. (1995). Defect in formation of functional matrix vesicles by growth plate chondrocytes in avian tibial dyschondroplasia: Evidence of defective tissue vascularization. *J Bone Miner Res* 10:1625-1634

Nudelman, F., Pieterse, K., George, A., Bomans, P.H., Friedrich, H., Brylka, L., Hilbers, P., de With, G. & Sommerdijk, N. (2010). The role of collagen in bone apatite formation in the presence of hydroxyapatite nucleation inhibitors. *Nat Mater* 9, 12: 1004-1009.

Olsen, B.R. (1989). The next frontier: molecular biology of extracellular matrix. *Connect Tissue Res* 23:115-121

Orimo, H. (2010). The mechanism of mineralization and the role of alkaline phosphatase in health and disease. *J Nippon Med Sch* 77: 4-12

Poole, A.R.; Matsui, Y.; Hinek, A. & Lee, E.R. (1989). Cartilage macromolecules and the calcification of cartilage matrix. *Anat Record* 224: 167-179

Pouget, E.M., Bomans, P.H., Goos, J.A., Frederik, P.M., de With, G. & Sommerdijk, M.A. (2009). The initial stages of template-controlled CaCO3 formation revealed by cryo-TEM. *Science* 323: 1455-1458.

Pourmand, E.P.; Binderman, I.; Doty, S.B.; Kudryashov, V. & Boskey, A.L. (2007). Chondrocyte apoptosis is not essential for cartilage calcification: Evidence from an in vitro avian model. *J Cell Biochem* 100:43-57

Pugliarello, M.C.; Vittur, F.; de Bernard, B.; Bonucci, E. & Ascenzi, A. (1970). Chemical modifications in osteones during calcification. *Calcif Tissue Res* 5:108-114

Rahman, M.A. & Oomori, T. (2010). In vitro regulation of CaCO(3) crystal growth by the highly acidic proteins of calcitic sclerites in soft coral, Sinularia Polydactyla. *Connect Tissue Res* 50:285-293

Ranvier, L. (1875-1882). *Traité Technique D'Histologie*. Librairie F. Savy, Paris

Register, T.; McLean, F.M.; Low, M.G. & Wuthier, R.E. (1986). Roles of alkaline phosphatase and labile internal mineral in matrix vesicle-mediated calcification. Effect of selective release of membrane-bound alkaline phosphatase and treatment with isosmotic pH 6 buffer. *J Biol Chem* 261:9354-9360

Register, T.; Warner, G.P. & Wuthier, R.E. (1984). Effect of L- and D-tetramisole on ^{32}Pi and ^{45}Ca uptake and mineralization by matrix vesicle-enriched fractions from chicken epiphyseal cartilage. *J Biol Chem* 259:922-928

Robison, R. (1923). The possible significance of hexosephosphoric esters in ossification. *Biochem J* 17: 286-293

Roughley, P.J. (2006). The structure and function of cartilage proteoglycans. *Europ Cells Mater* 12: 92-101

Sauer, G.R. & Wuthier, R.E. (1988). Fourier transform infrared characterization of mineral phases formed during induction of mineralization by collagenase-released matrix vesicles in vitro. *J Biol Chem* 263:13718-13724

Schaefer, L. & Schaefer, R.M. (2009). Proteoglycans: from structural compounds to signaling molecules. *Cell Tissue Res* 339: 237-246

Schäfer, EA. (1897). *A Course of Practical Histology* (second edition), Lea Brothers & Co., Philadelphia

Schäfer, EA. (1907). *The Essentials of Histology* (seventh edition), Lea Brothers & Co., Philadelphia and New York.

Scherft, J.P. & Moskalewski, S. (1984). The amount of proteoglycans in cartilage matrix and the onset of mineralization. *Metab Bone Dis Rel Res* 5: 195-203

Shepard, N. (1992). Role of proteoglycans in calcification, In.: *Calcification in biological systems*, Bonucci, E (ed.), pp. 41-58, CRC Press, Boca Raton

Silvestrini, G.; Zini, N.; Sabatelli, P.; Mocetti, P.; Maraldi, N.M. & Bonucci, E. (1996). Combined use of malachite green fixation and PLA2-gold complex technique to localize phospholipids in areas of early calcification of rat epiphyseal cartilage and bone. *Bone* 18:559-565

Simmer, J.P. & Hu, J.C. (2002). Expression, structure, and function of enamel proteinases. *Connect Tissue Res* 43:441-449

Sobel, A.E. (1955). Local factors in the mechanism of calcification. *Ann N Y Acad Sci* 60:713-731

Takagi, M. & Toda,Y. (1979). Electron microscopic study of the intercellular activity of alkaline phosphatase in rat epiphyseal cartilage. *J Electron Microsc* 2:117-127

Takagi, M.; Parmley, R.T. & Denys, F.R. (1983). Ultrastructural cytochemistry and immunocytochemistry of proteoglycans associated with epiphyseal cartilage calcification. *J Histochem Cytochem* 31:1089-1100

Takagi, M.; Parmley, R.T.; Denys, F.R.; Yagasaki, H. & Toda, Y. (1984). Ultrastructural cytochemistry of proteoglycans associated with calcification of shark cartilage. *Anat Record* 208:149-158

Takechi, M. & Itakura, C. (1995a). Ultrastructural and histochemical studies of the epiphyseal plate in normal chicks. *Anat Rec* 242:29-39.

Takechi, M. & Itakura, C. (1995b). Ultrastructural studies of the epiphyseal plate of chicks fed a vitamin D-deficient and low-calcium diet. *J Comp Pathol* 113:101-111

Takeuchi, A.; Ohtsuki, C.; Miyazaki, T.; Kamitakahara, M.; Ogata, S.; Yamazaki, M.; Furutani, Y.; Kinoshiya, H. & Tanihara, M. (2005). Heterogeneous nucleation of hydroxyapatite on protein: structural effect of silk sericin. *J R Soc Interface* 2:373-378

Thyberg, J.; Nilsson, S & Friberg, U. (1975). Electron microscopic and enzyme cytochemical studies on the guinea pig metaphysis with special reference to the lysosomal system of different cell types. *Cell Tissue Res* 156:273-299

van der Rest, M.; Rosenberg, L.C.; Olsen, B.R. & Poole, A.R. (1986). Chondrocalcin is identical with the C-propeptide of type II procollagen. *Biochem J* 237:923-925

Veis, A. (2003). Mineralization in organic matrix frameworks. *Rev Mineral Geochem* 54:249-289

Vittur, F.; Zanetti, M.;Stagni, N. & de Bernard, B. (1979). Further evidence for the participation of glycoproteins to the process of calcification. *Perspect Inherit Metab Dis* 2:13-30

Warner, G.P.; Hubbard, H.L.; Lloyd, G.C. & Wuthier, R.E. (1983). ^{32}Pi- and ^{45}Ca-metabolism by matrix vesicle-enriched microsomes prepared from chicken epiphyseal cartilage by isosmotic Percoll density-gradient fractionation. *Calcif Tissue Int* 35:327-338

Weiner, S. & Addadi, L. (1991). Acidic macromolecules of mineralized tissues: the controllers of crystal formation. *Trends Biochem Sci* 16: 252-256

Wheeler, E.J. & Lewis, D. (1977). An X-ray study of the paracrystalline nature of bone apatite. *Calcif Tissue Res* 24:243-248

Wu, L.N.Y.; Genge, B.R.; Dunkelberger, D.G.; LeGeros, R.Z.; Concannon, B. & Wuthier, R.E. (1997a). Physicochemical characterization of the nucleational core of matrix vesicles. *J Biol Chem* 272:4404-4411

Wu, L.N.Y.; Ishikawa, Y.; Sauer, G.R.; Genge, B.R.; Mwale, F.; Mishima, H. & Wuthier, R.E. (1995). Morphological and biochemical characterization of mineralizing primary cultures of avian growth plate chondrocytes: Evidence for cellular processing of Ca2+ and Pi prior to matrix mineralization. *J Cell Biochem* 57:218-237

Wu, L.N.Y.; Wuthier, M.G.; Genge, B.R. & Wuthier, R.E. (1997b). In situ levels of intracellular Ca^{2+} and pH in avian growth plate cartilage. *Clin Orthop* 335:310-324

Wu, L.N.Y.; Yoshimori, T.; Genge, B.R.; Sauer, G.R.; Kirsch, T.; Ishikawa, Y. & Wuthier, R.E. (1993). Characterization of the nucleational core complex responsible for mineral induction by growth plate cartilage matrix vesicles. *J Biol Chem* 268:25084-25094

Wuthier, R.E. (1992). Matrix vesicles: formation and function. Mechanisms in membrane/matrix-mediated mineralization, In: *Chemistry and biology of mineralized tissues,* Slavkin, H. & Price, P. (eds.), pp 143-152, Excerpta Medica, Amsterdam

Wuthier, R.E. & Lipscow, G.F. (2011). Matrix vesicles: structure, composition, formation and function in calcification. *Front Biosci* 1; 17: 2812-2902

Wuthier, R.E. & Register, T.C. (1985). Role of alkaline phosphatase, a polyfunctional enzyme, in mineralizing tissues, In: *The chemistry and biology of mineralized tissues,* Butler,W.T. (ed.), pp.113-124numerr linea, Ebsco Media, Inc., Birmingham

Part 3

Applied Biomineralization

6

Biomimetic Materials Synthesis from Ferritin-Related, Cage-Shaped Proteins

Pierpaolo Ceci[1], Veronica Morea[1], Manuela Fornara[1], Giuliano Bellapadrona[2], Elisabetta Falvo[3] and Andrea Ilari[1]
[1]*C.N.R. Institute of Molecular Biology and Pathology, Rome*
[2]*Department of Materials and Interfaces, Weizmann Institute of Science, Rehovot*
[3]*Regina Elena Cancer Institute, Pharmacokinetic/Pharmacogenomic Unit, Rome*
[1,3]*Italy*
[2]*Israel*

1. Introduction

Protein cages are characterized by a quaternary structure consisting in an assembly of multiple subunits endowed with highly similar three-dimensional structures. These assemblies enclose hollow spaces that can be used as ideal templates for the encapsulation of nano-material cargos. Indeed, the uniformity of the quaternary structure guarantees the attainment of nanoparticles (NPs) that are highly homogeneous in both size and shape, and the interior of the cage provides an isolated environment, shielded from bulk solution, where chemical reactions can take place. The protein surface is decorated by diverse chemical groups (*i.e.*, primary amines, carboxylates, thiols) that can be genetically and/or chemically manipulated in order to confer specific functionalities to the nano-cage. Further advantages of protein cages include their usually remarkable stability, which is generally higher than that of liposome-based molecules, and high solubility in water, as well as the fact that protein-encapsulated NPs can be produced in industrial strains on large-scale, high yield (grams or even kilograms) and low cost. Finally, the determination of the three-dimensional structures of protein-cages at the atomic level has provided relevant information to understand the biomineralization processes and design protein variants aimed at the creation of new functional biomaterials.

Members of the ferritin family, including ferritins and **D**NA-binding **p**roteins from **s**tarved cells (Dps), are among the most widely studied cage-like proteins. The unique template structures of these proteins have been used for the synthesis of a variety of new, non physiological mineral cores within the protein shell.

Ferritins are formed by 24 subunits assembled with octahedral (432) symmetry, whereas Dps are formed by 12 subunits assembled with a 23 rotational tetrahedral symmetry. Despite their different size and architecture, both ferritins and Dps assemblies have a spherical shape enclosing an inner cavity where iron is conserved in a non-toxic and bioavailable form. Both these molecules play a key role in cellular protection by maintaining iron homeostasis. In addition, Dps protect DNA from oxidative damage by a chemical mechanism, involving the consumption of hydrogen peroxide produced by the oxidative metabolism, and by a physical mechanism, consisting in DNA condensation and shielding. The shells of ferritins and Dps are traversed by pores allowing the passage of ions and small molecules. In particular, hydrophilic pores, lined with negatively charged residues, are involved in the uptake of Fe(II) ions, which are guided through the pores by a negative electrostatic charge (**Fig. 1**). After entering the pores, reduced iron atoms reach specific catalytic sites, named ferroxidase centres, where Fe(II) is oxidized to Fe(III) either by molecular oxygen (in ferritins) or hydrogen peroxide (in Dps). Fe(III) then moves to the protein cavity and binds to specific nucleation sites where it forms ferrihydrite (Fe(III)OOH) clusters, which grow until a single core (NP) of up to 500 iron atoms in Dps and 4000-5000 iron atoms in ferritins is formed.

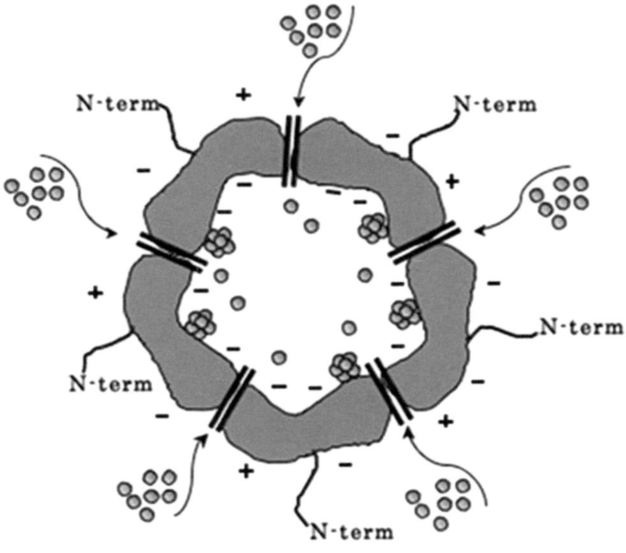

Fig. 1. Schematic illustration of ferritin. A negative electrostatic charge facilitates the passage of ions and small molecules through the pores traversing the protein shell. The quaternary structure consists of 24 subunits whose amino-termini (denoted with N-term) protrude from the external surface. Plus and minus signs indicate the higher concentration of negative charges on the internal surface with respect to the external one

A different mechanism of biomineralization is represented by the formation of specific intracellular structures called magnetosomes, which allow magnetotactic bacteria to navigate along geomagnetic fields. Magnetosomes originate from invaginations of the inner membrane and contain magnetite (iron oxide) or greigite (iron sulfide) nanoparticles that can reach dimensions up to 70 nm in diameter (Mann et al., 1990). Magnetosome assembly and function are governed by several proteins (Mam A, B, C, D, E) (Grünberg et al., 2001). Commercial uses of bacterial magnetosome particles have been suggested, including magnetic targeting of pharmaceuticals and enhancement of contrast agents in magnetic resonance imaging.

In this chapter we will describe the structural properties of ferritin-like proteins, the mechanisms of metal incorporation and the use of ferritin-like proteins as size-constrained reaction vessels. In particular, we will examine the main parameters involved in the biomineralization process, *i.e.*, the external/internal charge distribution and the number and type of metal binding/nucleation sites present inside the protein cavity. Finally, we will discuss the current and potential applications of ferritins in the fields of biomedicine, catalysis and electronics.

2. Structural and functional properties of cage-like proteins in biology: Ferritins and Dps

2.1 Iron homeostasis

Nearly all forms of life require iron. This metal commonly occurs in either the ferrous (Fe(II)) or ferric (Fe(III)) oxidation state. The wide range of the Fe(II)/Fe(III) redox potential (from approximately −500 to +600 mV, depending on the iron ligands and surrounding chemical environment) as well as the ability of iron to gain and lose electrons, thus cycling between the two oxidation states, makes iron an ideal redox catalyst in many biological processes. However, iron can also be a major problem for the cell. Firstly, Fe(III) is highly insoluble and precipitates in the cytoplasm as part of the uncharged species $Fe(OH)_3{\cdot}3(H_2O)$ (Ksp, 10^{-38}) (Aisen et al., 2001), thereby limiting the cellular concentration of free iron to 10^{-17} M at neutral pH. The reaction between Fe(II) and molecular oxygen is also a source of oxidative stress for the cell. The one electron reduction of Fe(II) by O_2 results in the formation of the superoxide radical $O_2^{\bullet-}$, which can accept another electron and two protons to produce hydrogen peroxide, H_2O_2. The superoxide radical and hydrogen peroxide are the by-products of incomplete O_2 reduction, and their balance is regulated by the enzyme superoxide dismutase (McCord & Fridovich, 1988). In addition, hydrogen peroxide can react again with Fe(II) generating hydroxyl radical species ($^{\bullet}OH$) via the Fenton reaction (1), while the $O_2^{\bullet-}$ radical can reduce ferric iron to ferrous ions (2):

$$Fe(II) + H_2O_2 \rightarrow Fe(III) + {}^{\bullet}OH + OH^- \tag{1}$$

$$O_2^{\bullet-} + Fe(III) \rightarrow O_2 + Fe(II) \tag{2}$$

The sum of these two reactions, named Haber–Weiss reaction (3), occurs in the presence of catalytic amounts of free iron ions and produces the hydroxyl anion, hydroxyl radical and O_2:

$$H_2O_2 + O_2^{\bullet-} \rightarrow OH^- + {}^{\bullet}OH + O_2 \tag{3}$$

The reactive oxygen species (ROS) involved in the Haber–Weiss reaction (*i.e.*, all the involved species with the exception of molecular oxygen) are highly toxic for the cell. In particular, hydroxyl radicals can oxidize biological macromolecules, such as DNA, proteins and lipids, and all organisms have developed strategies to store iron in a non toxic and readily bio-available form.

Iron bioavailability is extremely poor in aerobic organisms under physiologic conditions, such that iron is often a limiting nutrient for growth and virulence in bacterial pathogens (Andrews et al., 2003). In conditions of low iron availability, the metal is scavenged from the environment by efficient systems. Ferric chelators called siderophores are often secreted and internalized. Host-iron complexes, such as transferrin, lactoferrin, heme, haemoglobin, are directly used as iron sources by pathogens. Highly specific proteins transport iron complexes and heme across the cell wall of gram-positive bacteria and the outer membrane of gram-negative bacteria. ABC transporters carry heme and iron siderophores across the cytoplasmic membrane (Braun & Hantke, 2011).

One of the means that cells have adopted to protect themselves from the potentially toxic effects of free iron and radical chemistry is represented by proteins belonging to the ferritin family, including ferritins, bacterioferritins (Bfrs) and Dps proteins.

2.2 Ferritins

Ferritins are involved in iron storage and detoxification in most living organisms from microorganisms to plants, invertebrates, and mammals (Crichton & Boelaert, 2009). They are formed by 24 similar or identical subunits assembled to form a 24mer. A non-toxic, water-soluble, yet bioavailable iron core, often consisting in a ferric oxy-hydroxide mineral, is stored within the ferritin hollow shell.

Mammalian ferritins were the first to be studied from a structural and functional point of view. Typically, they are heteropolymers composed by two types of subunits named H (heavy, predominant in heart) and L (light, predominant in liver) (Drysdale, 1976). These chains are highly similar both in sequence (~55% sequence identity) and structure, since the L- and H-chain subunits can be super-imposed with an RMSD of 0.6 Å over ~170 amino acid residues (see Table 1). Additionally, they are isostructural, meaning that they can assemble in any proportion in the 24mer.

In prokaryotes there are two types of ferritins: bacterial ferritins, closely related to mammalian ferritins, and Bfrs, characterized by the presence of up to 12 heme groups per 24mer, which do not play a role in the iron oxidation and incorporation processes (Frolow et al., 1994). Sequence analyses have revealed that the sequence identity between ferritins and Bfrs is very low, generally below 15%, but the catalytically important residues are conserved (Le Brun et al., 2010). The structural analyses discussed below demonstrate that, despite their low degree of sequence identity, ferritins and Bfrs conserve the structural features required for iron oxidation and incorporation.

2.3 Dps proteins

Dps proteins, whose prototype was discovered in *Escherichia coli* (Almiron et al., 1992), are found exclusively in prokaryotes, where they are expressed under starvation and oxidative stress conditions. Like ferritins, Dps possess iron storage and detoxification capabilities, and the three-dimensional structure of individual Dps protein subunits is very similar to that of ferritin subunits. However, their quaternary assemblies are different, in that they comprise only 12 identical subunits assembled with 23 rotational tetrahedral symmetry. Therefore, the hollow shell enclosed by Dps has a smaller volume than the ferritin one, and incorporates a 10-fold smaller number of iron atoms.

Additionally, differences between Dps proteins and ferritins concern the ferroxidase centre. This is entirely comprised within each subunit in ferritin and Bfrs, whereas it is contributed by two-fold symmetry related subunits in Dps proteins (Ilari et al., 2000). Moreover, in Dps proteins it catalyzes the oxidation of ferrous iron by hydrogen peroxide within the protein shell, thus preventing the diffusion of ROS in the cytoplasm and, consequently, damage of cellular components (Bellapadrona et al., 2010).

In addition to this chemical mechanism of protection, some Dps proteins are able to protect DNA from oxidative damage by binding DNA, thus providing it with a physical shield against the damage mediated by reactive oxygen species (ROS). Dps proteins have been shown to bind DNA by one of three mechanisms, based on the following structural features: i) presence of a long N-terminal tail rich in positively charged residues, which protrudes from the dodecameric structure, as in the prototypic Dps from *E. coli* (Grant et al., 1998); ii) presence of a C-terminal tail, also protruding from the structure and containing positively charged residues, as in *M. smegmatis* Dps; iii) presence of a positively charged protein surface, as in the Dps from *H. pylori*. The physical DNA protection activity is largely influenced by the net charge and charge distribution on the protein surface, and it has been shown to be dependent on both pH and salt concentration (Chiancone & Ceci, 2010).

2.4 Building blocks: Structural monomeric units

Comparisons between the three-dimensional structures of the H and L chains of mammalian ferritins, bacterial ferritins and Bfrs subunits have revealed a striking similarity of the monomer subunit (Crichton & Declercq, 2010).

Monomers of all these proteins are folded in a characteristic four-helical bundle, formed by four antiparallel helices (A-D), and a shorter helix on the top of them (E) **(Fig. 2A)**. The most variable part of the monomer fold is represented by the loops connecting the helices, in particular the long BC loop, which stretches along the length of the helical bundle **(Fig. 2A)**, and by the short non-helical regions at the N- and C-termini. The relative orientation of the E and D helices varies in different proteins. They form an angle of 90° in both bacterial ferritins and heme-containing Bfrs, and of about 60° in mammalian ferritins. In the quaternary structure the N-terminus, BC loop, and helices A and C face the external side of the spherical shell, while helices B and D are on the internal side.

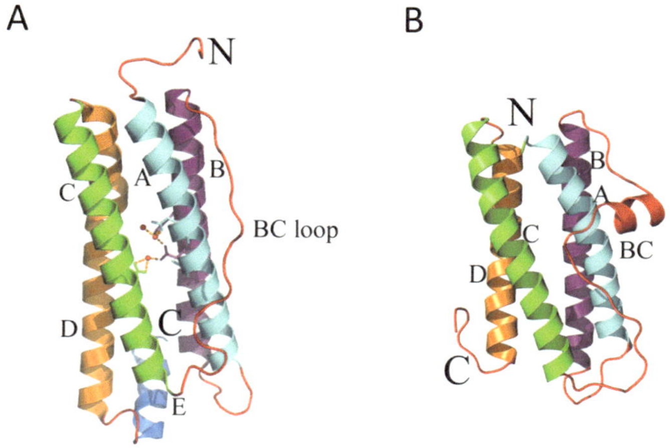

Fig. 2. Monomeric fold of ferritins and Dps. A. Human H ferritin (PDB code: 2CEI). Helices A-D forming the four-helical bundle, the E-helix and the BC loop are indicated. B. Dps from *Listeria innocua* (PDB code: 1QGH). Helices A-D forming the four-helical bundle and the BC-helix are indicated. This picture has been generated with PyMol (Delano Scientific LLC, San Carlos, LA; http://www.pymol.org).

The Dps monomer also folds into a four-antiparallel-helix bundle (**Fig. 2B**), which is almost entirely super-imposable to that of ferritins (Table 1) and Bfrs. The four helices A–D are stabilized mainly by inter-helical hydrophobic interactions. Helices B and C are connected by a long BC loop, which comprises the short BC helix. Helix E, present in the ferritin C-terminal region, is absent in Dps.

2.5 Functional components: Ferroxidase centres

Mammalian H chain ferritins are characterized by the presence of a di-iron ferroxidase centre, involved in the oxidation of Fe(II) to Fe(III). This is not present in L chains, which contain a group of conserved negatively charged residues (Glu57, Glu60, Glu61, Glu64 and Glu67, mouse L-chain numbering) directed into the inner cavity of the protein and known to be involved in the ferrihydrite nucleation process (Andrews, 2010; Drysdale, 1976; Granier et al., 2003; Nordlund & Eklund 2006).

The ferroxidase centre of H-chain ferritins has been intensively studied for several decades. However, due to instability of the iron bound form a structure of their ferroxidase centre in complex with iron is not available yet. Nevertheless, structural data on H-chains in complex with different metals, like Zn(II) and Tb(III), have been determined, based on which detailed studies on the bimetal binding centre have been performed (Lawson et al., 1989; Toussaint et al., 2007). The ferroxidase centre of mammalian H chain ferritins comprises two iron binding sites, A and B. The iron atom at site A (FeA) is ligated by one histidine, one monodentate glutamate and one bridging glutamate (His65, Glu27 and Glu62, human H chain numbering; **Fig. 3A**) (Santambrogio et al., 1996; Trikha et al., 1995). The iron atom at site B (FeB) is ligated by Glu62 and by two glutamate residues (Glu61 and Glu107). The conservation of the amino acid residues belonging to the ferritin ferroxidase site is highlighted by sequence alignments (Andrews et al., 1992). Although mutagenesis studies have shown that both sites A and B are important for iron uptake and oxidation, only a few ferritin structures have been determined with both sites occupied by metals. The structure of the H chain homopolymer in complex with Tb(III) revealed that Glu61 is able to adopt two conformations (Lawson et al., 1989). In one conformation the residue is bound to the metal at the B site and in the other it is projected towards the internal cavity (**Fig. 3A**).

The chemical environment of the ferroxidase centres of representative bacterial ferritins and Bfrs is similar to that observed in the crystal structure of recombinant human H chain ferritin (HuHFt) in complex with Zn(II) (Stillman et al., 2001). As shown in **Fig. 3B** and **3C** the ferroxidase centres of bacterial ferritins and Bfrs present many similarities with that of HuHFt. In all the ferritin monomers the iron atom in the A site is coordinated by one histidine and one glutamate residue. The iron atom in the B site is ligated by two glutamate residues and one glutamate residue bridges both metal centres.

The structures of bacterial ferritins display two main differences with respect to that of HuHFt. In *P. furiosus* ferritin (Tatur et al., 2007), whose crystals had been soaked with Fe(II), a third site (site C) has been found to be occupied by an iron atom (**Fig. 3B**) and the iron at the B site is coordinated by Glu130 as an additional ligand.

The ferroxidase centre of Bfrs displays major differences with respect to that of HuHFt and bacterial ferritins. Although the structure of the first Bfr was reported in 1994, the first iron-bound structure (*D. desulfuricans* Bfr) was described several years later (Romão et al., 2000).

Fig. 3C depicts the ferroxidase centre of the Bfr from *E. coli* (Crow et al., 2009), similar to that of Bfr from *D. desulfuricans*, which was obtained by soaking the protein crystals with a solution containing Fe(II). In this structure, the water molecule coordinating FeA in HuHFt is substituted by Glu127, which bridges the metal sites A and B. Moreover, FeB is coordinated by His130, which substitutes Glu130 present in bacterial ferritins.

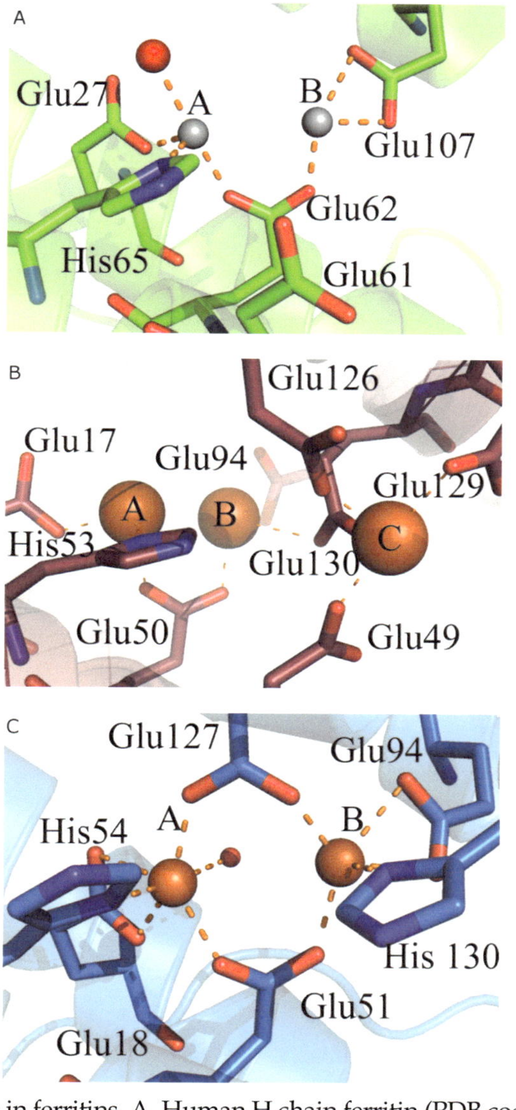

Fig. 3. Ferroxidase centre in ferritins. A. Human H chain ferritin (PDB code: 2CEI). B. *Pyrococcus furiosus* bacterial ferritin (PDB code: 2JD7). C. *E. coli* Bfr (PDB code: 3E1M). Residues are shown as sticks and coloured according to atom type: N, blue; O, red; C, green, salmon and light blue in panels A, B and C, respectively. Zn(II) atoms and the oxygen atom of a coordinated water molecule in panel A, and Fe(III) atoms in panels B and C are shown as spheres and coloured grey, red and orange, respectively. The picture was generated using PyMol

In all known ferritins and Bfrs the ferroxidase centres are embedded within each of the four-helical bundle monomers. Conversely, the functional centres of Dps proteins are located at the interface of each pair of 2-fold symmetry-related subunits (**Fig. 4A**), both of which provide the iron ligands. In particular, in the X-ray crystal structure of *L. innocua* Dps (Ilari et al., 2000), two histidine residues (H31 and H43) are provided by one subunit and two carboxylate ligands (D58 and E62) by the symmetry-related subunit within the dimer (**Fig. 4B**). These ferroxidase centre ligands are conserved in all Dps with the only known exception of *T. elongatus* DpsA, where His78 replaces the canonical aspartate (D58, *Listeria* numbering) (Alaleona et al., 2010).

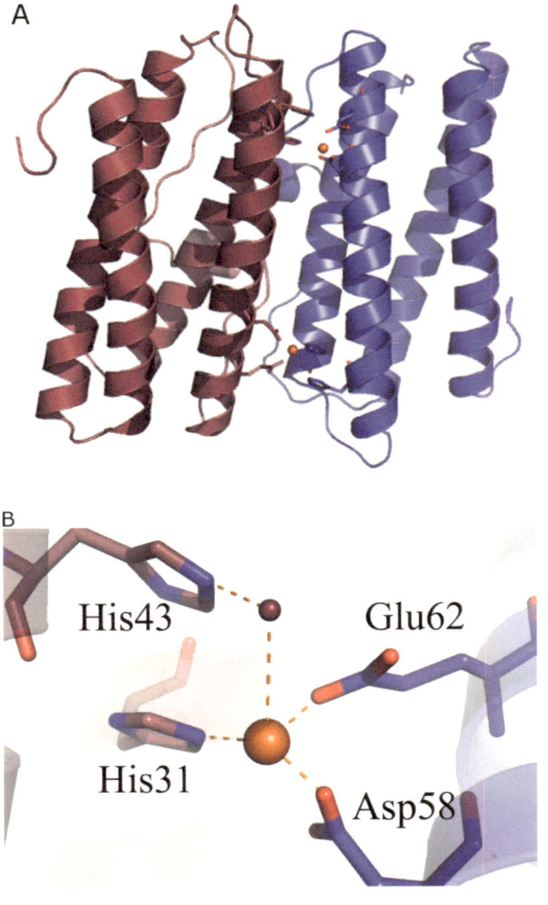

Fig. 4. Ferroxidase centre in Dps proteins. A. Two ferroxidase centres are present at the dimeric interface of the Dps from *Listeria innocua* (*Li*Dps) (PDB code: 1QGH). The 2-fold symmetry-related subunits are coloured dark red and blue, respectively. B. Blow up of one ferroxidase centre of *Li*Dps. Iron within the A site and the water molecule at the B site are indicated as spheres and coloured orange and red, respectively. Residues involved in iron or water coordination are shown as sticks and coloured according to atom type: N, blue; O, red; C, dark red (histidines from one subunit) or blue (acidic residues from the symmetry-related subunit within the dimer).

The occupancy of the two metal binding sites with iron varies significantly in the known crystal structures. In *L. innocua* Dps the ferroxidase centre contains one iron and one water molecule, which lies at about 3 Å from the iron. It has been proposed that this water molecule may be replaced by a second iron atom to give rise to a canonical bimetallic ferroxidase centre (Ilari et. al., 2000). In *D. radiodurans* Dps2 and *B. brevis* Dps, both sites are occupied by Fe(III) (Cuypers et al., 2007; Ren et al., 2003). In *H. pylori* Dps, *D. radiodurans* Dps1 and *A. tumefaciens* Dps structures, only the A site contains iron whereas the B site contains a water molecule (Ceci et al., 2003; Romão et al., 2006; Zanotti et al., 2002). In *E. coli* Dps (Grant et al., 1998) and *T. elongatus* Dps both sites are occupied by two water molecules (Franceschini et al., 2006). These data indicate that the two metal binding sites A and B have different affinity for iron, and that the second metal coordination shell, formed by residues that are not directly in contact with the metal but interact with metal binding site residues, may influence the affinity for iron. As an example, the low affinity of *E. coli* Dps for iron has been ascribed to the presence of Lys48 in the second metal coordination shell, which forms a salt bridge with Asp78 in the ferroxidase centre, reducing the iron coordination propensity of this residue (Ilari et al., 2002).

All Dps proteins bind and oxidize iron at the ferroxidase centre. The only known exception is represented by DpsA from *T. elongatus*. The peculiar set of iron coordination residues of this protein, where the conserved Asp58 (*Listeria* numbering) is replaced by His78, endows the A site with a strong affinity for Zn(II) (Alaleona et al., 2010).

2.6 Complete structure: Quaternary assembly

Ferritins are formed by 24 subunits assembled to form a compact, symmetric and extremely stable apoferritin shell. This shell has the approximate geometry of an octahedron and, therefore, possesses three four-fold axes passing through the centre of two opposite faces (**Fig. 5A**), four three-fold axes passing through two opposite vertices (**Fig. 5B**) and six two-fold axes passing through the centre of two opposite edges (**Fig. 5C**).

Vertebrate ferritins present two hydrophilic pore entrances along each three-fold axis, which allow Fe(II) to enter inside the protein. These 8 pore entrances have a ~ 4 Å diameter and are lined by Asp131 and Glu134 (HuHFt numbering) from each of the three-fold symmetry related subunits (Bou-Abdallah et al., 2008). Both the mutation of these negatively charged residues and the presence of Tb(III) or Zn(II) decrease the capacity of ferritins to oxidate and incorporate iron ions (Bou-Abdallah et al., 2003; Watt et al.,1988). The pores stretching along the four-fold symmetry axes are mostly hydrophobic, with the six pore entrances lined by leucine residues. These are believed to be the major entrance route for molecular oxygen and hydrogen peroxide inside the ferritin cavity (Liu & Theil, 2005).

With respect to vertebrate ferritins, in *E. coli* ferritin and Bfr (Stillman et al., 2001), *Azotobacter vinelandii* Bfr (Liu et al., 2004) and in ferritins from the archaeal hyperthermophiles and anaerobes *Pyrococcus furiosus* (Tatur et al., 2007) and *Archaeoglobus fulgidus* (Johnson et al., 2005), the three-fold channels are less hydrophilic while the 4-fold channels are more polar and considered to be a likely route of iron entry and release.

Identical Dps subunits give rise to dodecameric shell-like assemblages characterized by 23 rotational tetrahedral symmetry. Accordingly, the Dps molecule is smaller than ferritin and this difference in size is reflected in a lower capacity to store iron (~500 versus ~4500 atoms per oligomer).

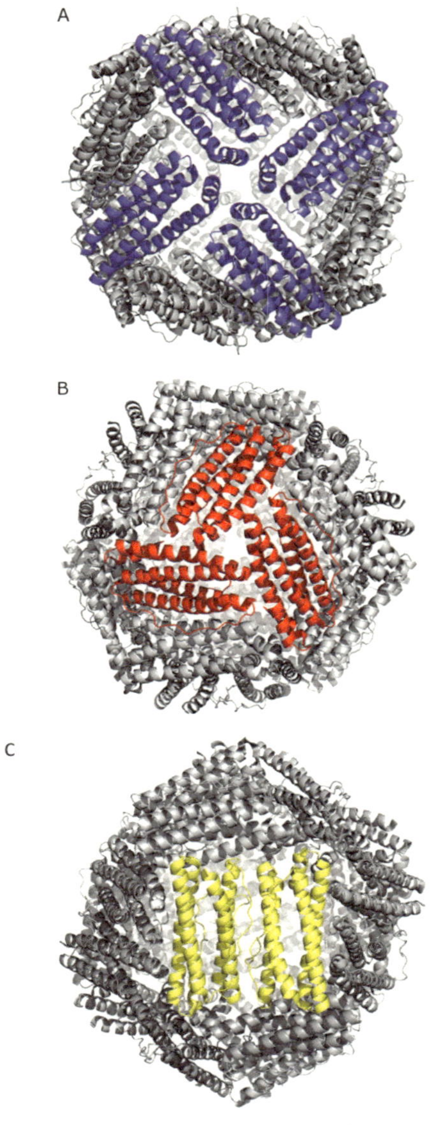

Fig. 5. Quaternary assembly of ferritin from *Pyrococcus furiosus* (PDB code: 2JD7). A. View along the four-fold axis. B. View along the three-fold axis. C. View along the two-fold axis. The subunits forming the four-, three-, and two-fold interfaces are indicated in blue, red and yellow, respectively, the other subunits are grey.

The Dps dodecamer can be viewed as an ensemble of four trimers placed at the vertices of a tetrahedron. There are three two-fold axes passing through the centres of the tetrahedron edge and the centre of the structure, and four three-fold axes passing through the vertices of the tetrahedron and the centres of the opposite face. Consequently, there are two non-equivalent environments along the three-fold axes, which have been designated "ferritin-like" (**Fig. 6A**) and "Dps-type" (**Fig. 6B**), respectively (Ilari et al., 2000). The ferritin-like interfaces resemble

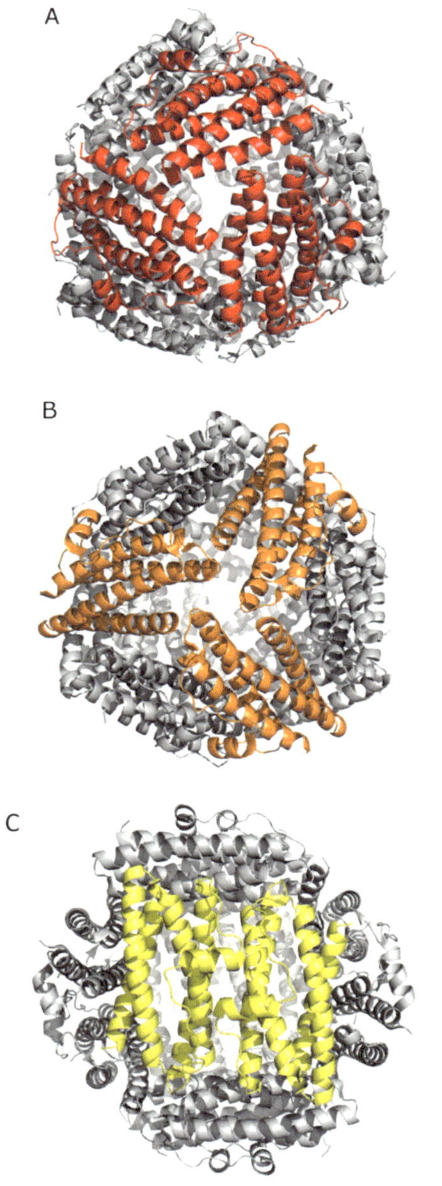

Fig. 6. Quaternary assembly of Dps from *L. innocua* (PDB code: 1QHG). A. Front-view along the three-fold axis: ferritin-like interface. B. Back-view along the three-fold axis: Dps-like interface. C. View along the two-fold axis. The monomers forming the ferritin-like and Dps-like interfaces are indicated in red and orange respectively, the subunits forming the two-fold interface are yellow and the other subunits are grey.

those formed along the three-fold symmetry axes of the ferritin oligomer. Like the ferritin pores, ferritin-like pores are rich in negatively charged aspartate and glutamate residues. These amino acid residues are the major contributors to the negative electrostatic gradient driving iron within the protein cavity. The Dps-type interfaces are typical of these proteins and are formed by the C-terminal portions of the B and D helices (**Fig. 2A**). The Dps-type pores (external diameter 4.5 Å, **Fig. 6B**) are characterized by a marked variability in their chemical nature. The dimeric interface, comprising the ferroxidase centre, is stabilized by hydrophobic interactions involving mostly residues belonging to the short BC helices (**Fig. 6C**). The large surface area buried upon dimerization (about 1290 $Å^2$ per monomer) has been proposed to account for the stability of the dodecameric *Listeria innocua* Dps assemblage at high temperatures (≥ 70 °C) and over a wide pH range (Chiaraluce et al., 2000). The external surface of Dps proteins is generally rich in negatively charged residues, with the exception of the *H. pylori* protein, whose positively charged surface is considered to be responsible of the ability of this protein to bind DNA (Zanotti et al., 2002).

3. Iron mineralization in ferritins and Dps

3.1 The iron mineral core

Both ferritin and Dps proteins can accommodate polynuclear iron micelles within their internal cavities. Two different core types have been described: a native core, which is present in the "as purified" protein; and an *in vitro* core, which is formed after addition of iron to the apoprotein. Mammalian ferritins, as isolated, generally contain 1000-3000 atoms of iron per molecule, while Bfrs usually contain only 800-1500 iron atoms (Lewin et al., 2005). As described above, in ferritins the negatively charged channels along the 3-fold symmetry axes allow the entry of Fe(II); the ferroxidase sites located within the four-helix bundle of the so-called H-type subunits catalyze iron oxidation by O_2; and the negatively charged internal cavity allows deposition of an oxy-hydroxide core containing up to 4500 iron atoms by addition of iron to the apo- form.

A notable feature of the apoferritin moiety is its capacity to direct iron deposition towards the formation of microcrystalline structures. The characteristics of the structures formed by horse spleen and human ferritins have been elucidated by both Mossbauer spectroscopy and X-ray diffraction studies. Basically, the native microcrystalline core contains oxygen atoms packed 3 Å apart (Massover & Cowley, 1973) and iron atoms placed in spaces between oxygen layers, so that they can be coordinated either octahedrically or tetrahedrically. Although the native mineral core of eukaryotic ferritins has been traditionally described as being formed by ferrihydrite, more recent studies have revealed a polyphasic structure consisting of both ferrihydrite and other phases, including a magnetite-like phase. Ferrihydrite is less crystalline and, therefore, less stable, than other iron mineral phases. However, this phase is stable when prepared inside ferritin, highlighting the ability of biomolecules to direct and selectively stabilize specific polymorphs.

Structural information at atomic resolution for ferritin cores is lacking because preparations of iron-containing ferritins are generally poly-dispersed with respect to their metal components, even though their protein components are highly homogeneous. This has prevented determination of the structure of the mineral core within ferritins by X-ray crystallography. Poly-dispersion of the mineral core is probably due to the fact that the iron mineral can be potentially nucleated at 24 different positions relative to the crystal itself, leading to a smearing of electron density. However, other techniques that are frequently used in materials sciences have been used to clarify structural and chemical aspects of ferritin cores. Galvez et al (2008) have used Transmission Electron Microscopy (TEM), X-ray Absorption Near Edge Spectroscopy (XANES), Electron Energy-Loss Spectroscopy (EELS), Small-Angle X-ray Scattering (SAXS), and SQUID magnetic measurements to demonstrate that the structure of the iron core of horse ferritin is formed by several iron oxide phases (ferrihydrite, magnetite, hematite), whose relative percentages vary as iron is gradually removed from the metal core. In particular, the native ferritin core mainly corresponds to a ferrihydrite phase with physiological values of 1000-2000 iron atoms, while magnetite appears to be the predominant phase when the iron content of ferritin decreases below 500 atoms. Moreover, it has been shown that the size of the iron core does not vary significantly with iron removal because iron is removed from the more chemically labile ferrihydrite core, hollowing it out, rather than from the magnetite shell.

The iron mineral cores of prokaryotic ferritins display significant variability. Bfrs cores usually have higher phosphate content with respect to eukaryotic ferritins, and appear to be largely disordered as judged by TEM analyses. X-ray absorption fine structure (EXAFS) experiments on *A. vinelandii* Bfrs with a Fe:P ratio of about 1.7:1 have shown that phosphate groups reduce the number of near-iron neighbours thus explaining the decrease in the metal core order and supporting a model for the core in which phosphate is clearly an integral constituent (Rohrer et al., 1990). Native bacterial ferritin cores have not been investigated in detail, but studies of ferritin from *H. pylori* have revealed that its core contains significant amounts of iron and phosphate with a small amount of iron oxy-hydroxide, leading to the conclusion that it has a Bfrs-like core structure (Doig et al., 1993).

The native iron core of Dps-like proteins generally contains 5-50 iron atoms, with variable proportions of phosphate (Kauko et al., 2006; Stefanini et al., 1999; Yamamoto et al., 2002). The Dps core can also contain other metals, such as Zn, Cu, Cr, Mn, Co, Ni and Mo (Yamamoto et al., 2002). *In vitro*, Dps proteins can store a maximum of approximately 450 Fe atoms per dodecamer, an amount consistent with a protein cage significantly smaller than that of mammalian apoferritin. Iron-loaded *E. coli* Dps crystals have been studied using polarized single crystal absorption micro-spectrophotometry. In these crystals iron ions are oriented with tetrahedral symmetry, with the tetrahedron centre occupied by iron ions and the vertices by oxygen atoms (Ilari et al., 2002).

Different iron nanominerals with specific magnetic properties have been created using ferritin templates, and similar approaches may be envisaged for Dps proteins as well. The ferrimagnetic iron oxide phase Fe_3O_4 has been synthesized within ferritin proteins under nitrogen flux at temperatures above 60°C, using H_2O_2 as an additional oxidant agent. These materials have been extensively characterized and can be exploited in different applications (see below).

3.2 Iron mineralization mechanisms

The most widely accepted mechanism by which iron is mineralized in ferritins is shown in **Figure 7**, although alternative mechanisms have been proposed to occur and most mineralization studies have been performed *in vitro*, such that their conclusions may not necessarily apply to the *in vivo* process. According to the mechanism in **Fig. 7**, Fe(II) ions enter the ferritin molecule through the channels at the three-fold axes, which are known to have metal ion binding sites lined with carboxyl groups (Macara et al., 1973; Stefanini et al., 1989; Treffry et al., 1993; Wardeska et al., 1986). Ferritins molecules endowed with the catalytic ferroxidase activity (*i.e.*, vertebrate H chains and bacterial ferritins), rapidly oxidize Fe(II) ions to Fe(III) using O_2 or H_2O_2 as an oxidant (Watt et al., 1988; Xu & Chasteen, 1991; Yang & Chasteen, 1999). Further processing of Fe(III) includes movement to the hollow cavity and formation of the mineral core. Ion movements from the ferroxidase site towards the catalytic centres or the protein cavity is driven by an electrostatic potential (Douglas & Ripoll, 1998).

In vertebrate ferritins, L chain subunits are characterized by the presence of five negatively charged Glu residues (Glu57, Glu60, Glu61, Glu64 and Glu67, mouse L-chain numbering), located on the interior surface. These constitute the so-called nucleation sites and are mainly responsible for mineral core formation (Santambrogio et al., 1996). Bacterial ferritins and vertebrate H chains contain some Glu residues equivalent to those in the L chains, which have been suggested to be involved in the nucleation process as well (Bou-Abdallah et al., 2004; Lawson et al., 1991). Overall, the presence of carboxylate groups in both L and H chains is considered to be vital for efficient iron mineralization. As iron incorporation proceeds, Fe(II) oxidation reaction can take place also on the mineral surface formed inside the protein (**Fig.** 7). This reaction becomes predominant as the amount of Fe(II) atoms increases, and when the core is already formed (Bunker et al. 2005).

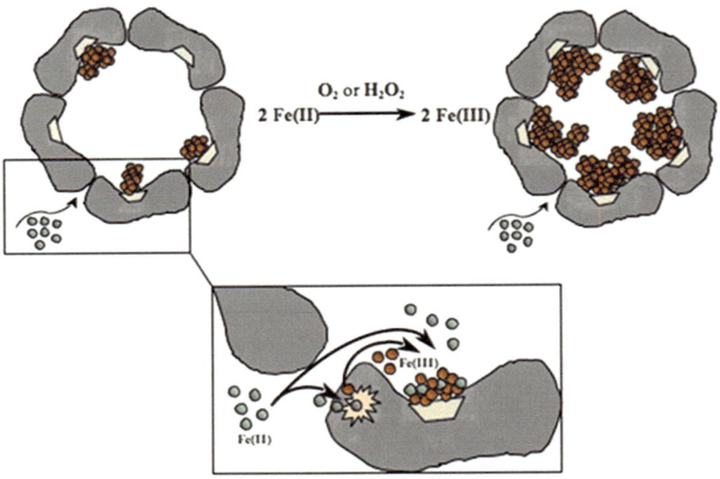

Fig. 7. A schematic illustration of the most widely accepted mechanism for iron incorporation in ferritin proteins. Details are described in the text. Briefly, Fe(II) atoms (green) enter the cavity enclosed by protein subunits (dark grey) *via* the hydrophilic pores. From these pores, Fe(II) atoms are driven to the ferroxidase centre (light pink) where they are oxidized to Fe(III). Fe(III) atoms (brown) move to the iron nucleation sites (light grey), where Fe(III)-mineral formation is initiated. When the Fe(III)-mineral reaches a sufficient size, Fe(II) atoms can also get oxidized directly on the surface of the growing mineral.

As in ferritins, in Dps proteins, after oxidation, iron moves to the protein internal cavity and gives rise to the mineral core. At variance with the ferroxidation step, iron uptake has similar characteristics in both ferritins and Dps proteins. The nucleation step takes place in the internal cavity and is likely to involve specific negative residues (Asp and Glu) lining the Dps internal cavity. Slow nucleation is followed by fast, cooperative growth of the crystal, as shown by the sigmoid kinetics of core formation (Ceci et al., 2010). Cooperativity in the crystal growth step is modulated by different factors, such as the strength of the electrostatic gradient at the pores, the nature of the oxidant used and ionic strength, which also determine the size distribution of the iron mineral (Bellapadrona et al., 2009; Ceci et al., 2010).

4. Biomimetic synthesis of nanoparticles in ferritins and Dps

NPs of several metal compounds, different from physiological mineral cores and generated using different approaches inside apoferritin or Dps proteins, have been described. These include Fe_3O_4, Co_3O_4, Mn_3O_4, Pt, CoPt, $Cr(OH)_3$, TiO_2, CdS, PbS, CdSe, ZnSe, $CaCO_3$, $SrCO_3$, $BaCO_3$, Pd, Cu, Ag or Au (Bode et al., 2011; Uchida et al., 2010).

Generally, the method used for NPs formation is based simply on the addition of various ions or compounds (simultaneously or sequentially) to apoferritin under specific conditions, allowing the diffusion of the ions or compounds inside the protein shell, followed by an oxidation (using O_2 or H_2O_2 as oxidant) or reduction (generally with $NaBH_4$) step (**Fig. 8**). It is important to notice that most of the mineralization reactions that have been successfully carried out are not specific to iron, suggesting that the electrostatic character of the interior surface of the protein cage is the major player in the mineralization process. Besides the presence of an electrostatic gradient that favours the concentration of cations inside the cavity, it is of great value to have multiple metal binding/nucleation sites able to incorporate efficiently the required metal, which are likely to act as starting points and metal particle seeds for the mineralization reaction (Kasyutich et al., 2010; Ueno et al., 2009). Other parameters have to be taken into account for the preparation of protein-enclosed nanomaterials of biotechnological value, such as: the specific ferritin molecule to be used as a template; the oxidant/reducing agent; and the reaction conditions (*i.e.*, ionic strength, pH, temperature, presence of chelating molecules, etc.).

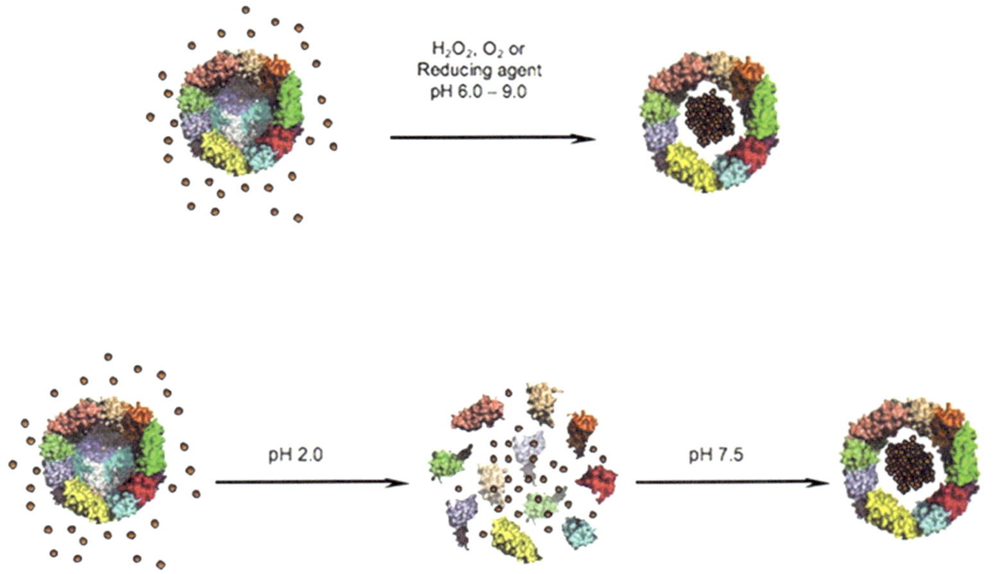

Fig. 8. Schematic illustration of nanoparticles synthesis using ferritins as nano-reactors (see text for details). Top. Ferritin-loading approach. Down. Assembly-disassembly approach.

An alternative approach from the general method described above has been used to incorporate metal particles and other compounds that cannot pass through the ferritin channels. The protein cage has been disassembled in extremely acidic conditions (pH ~2.0) and the desired compounds have been passively encapsulated within the cavity by raising the pH to neutral values (**Fig. 8**) (Yang et al., 2007).

Some of the recent and most significant publications on inorganic nanomaterial synthesis inside the apoferritin cavity are described below, especially in terms of their potential applications in different fields.

4.1 Biomedical applications

Imaging agents are among the compounds that can be successfully loaded within the interior cavity of ferritin proteins. Indeed, Douglas and co-workers have been able to synthesize magnetite NPs encapsulated within the internal cavity of recombinant HuHFt, which possessed T2* MRI properties comparing favourably with known iron oxide MRI contrast agents (*e.g.*, USPIO; Uchida et al., 2008). The same group has hypothesized that ferritins, which often accumulate in human plaques, may serve as an intrinsic vehicle for targeting plaque macrophages (Li et al., 2008), and has demonstrated that modified ferritin cages can be used as fluorescence or MRI agents for *in vivo* detection of vascular macrophages (Terashima et al., 2011).

Two other groups from different laboratories, namely the groups of Aime and of Domınguez-Vera, have prepared water-soluble gadolinium NPs with NMR longitudinal and transverse relaxivities higher than the ones of clinically approved paramagnetic Gd-chelates, thus indicating the great appeal of these novel classes of MRI contrast agents (Geninatti et al., 2006; Sanchez et al., 2009).

A different approach, in terms of the nature of tracer to be used for medical imaging, has been reported (Joh et al., 2011; Lin et al., 2011). Human ferritin has been loaded with radioactive metal ions (^{64}Cu). The ferritin nanotracer thus obtained possessed positron emission tomography (PET) functionalities for high sensitive tumor imaging.

Another potential application of ferritin in the biomedical field has been proposed by Babincová et al. (2000), who suggested to exploit the magnetic properties of the ferritin iron core for magnetic fluid hyperthermia (MFH). MFH is a promising new cancer treatment aimed at burning tumour cells. The procedure has been successfully used in glioma, prostate, liver, and breast tumours. Magnetic NPs should be applied directly to the tumour or injected into the body intravenously and diffuse selectively into cancerous tissues. Adding a safe high-frequency magnetic field (100-400 kHz), leads the particles to heat up, raising the temperature of the tumour cells without damaging the normal ones. However, so far data showing the heating capacity of super-paramagnetic cores encapsulated in ferritins are lacking.

Loading of metal-based drugs inside the ferritin cavity is another appealing opportunity for future tumour therapies. Recently, Xing et al. (2009) successfully encapsulated platinum-based anticancer drugs in the cavity of horse spleen ferritin. The loading capacity of ferritin differed with respect to different drugs but the protein shell remained intact after encapsulation, suggesting that this may be an alternative strategy for the delivery of platinum drugs.

For all the applications described so far, the development of molecules endowed with the ability to specifically target NPs to selected cells and tissues would be of great value. In this direction, the exterior surface of the ferritin assembly possesses all the features necessary to operate as an appropriate platform for specific cell targeting/delivery. In fact, modification of the exterior surface can be achieved either chemically or genetically. For instance, short peptide sequences and full length antibodies or their fragments, able to recognize specific cell receptors, have been genetically conjugated with the N-terminal region of HuHFt (Uchida et al., 2006 and our unpublished results), or chemically linked to reactive groups on the protein surface (Hainfeld, 1992 and our unpublished results). These findings suggest that cell/tissue specific delivery of imaging and therapeutic agents may be achieved by engineering chemical groups present on the exterior surface of ferritin proteins. Thus, NPs enclosed within HuHFt derivatized with targeting moieties may offer selective delivery possibilities *in vivo*, being presumably non-toxic and biocompatible.

4.2 Catalytic applications

The ferritin cage can serve as a catalytic nanoreactor for chemical reactions promoted by various metal catalysts. In this respect, the spatially restricted inner cavity of ferritin proteins can be exploited as an ideal chemical reaction chamber.

The group of Trevor Douglas has been among the first to investigate the use of ferritin-enclosed iron NPs as photoreduction catalysts (Kim et al., 2002). In particular, the native mineral core encapsulated within the protein cavity, mainly in the ferrihydrite form, has been shown to act as a semiconductor photocatalyst for the reduction of the highly toxic Cr(VI) to the more benign Cr(III). This work has been followed by several papers describing the use of ferritins or Dps proteins, loaded with different metals, in catalytic reactions (Ueno et al., 2004; Suzuki et al., 2009; Prastaro et al., 2009, 2010). The metal catalysts used in the reported organic reactions were generally based on palladium or gold atoms. Chemical reactions such as olefin hydrogenation, Suzuki-Miyaura cross-coupling followed by an enantioselective enzyme-catalyzed reduction to form chiral biaryl alcohols, and homocoupling of boronic acids and potassium aryltrifluoroborates have been described.

4.3 Electronic applications

Ferritin proteins have great potential as building blocks in the fabrication of electronic nanodevices. Yamashita's group has investigated the use of ferritin and Dps-enclosed metal NPs as building units, and has developed the so-called bio-nano process (BNP) for the fabrication of metal oxide semiconductors (MOS), such as floating nanodot gate memory devices or low-temperature polycrystalline silicon thin film transistor flash memories (Hikono et al., 2006; Ichikawa et al., 2007; Miura et al., 2006; Yamashita, 2008). In the BNP ferritin proteins are used as scaffolds to fabricate inorganic nanostructures on structures that are produced by conventional top-down method, such as photolithography. This combination of biological nanofabrication and top-down methods is an attractive process to produce nano-electronic devices. Recently, scientists from the same group have demonstrated that BNP has an advantage in the control of parameters like size, shape, and density of nanodot arrays of MOS devices (Yamada, et al., 2007).

The ability to control the magnetic properties of synthesized NPs is of high importance in the fabrication of magnetic devices on the nanoscale and in their applications. As mentioned above, proteins belonging to the ferritin family provide both a size and shape constrained reaction environment, which allows the magnetic properties of the synthesized magnetic

NPs to be tailored. For example, changing the size of the protein cage (*e.g.*, by switching from ferritins to Dps proteins, or *viceversa*) or the metal loading factor will produce NPs with different size, which might possess different magnetic properties (Gilmore et al., 2005; Klem et al., 2007).

Another approach to control the magnetic properties of ferritin-enclosed NPs consists in the creation of high-order structures based on ferritin cages (Kostiainen et al., 2011). Recombinant ferritins encapsulating Fe_3O_4−γ-Fe_2O_3 iron oxide (magnetoferritin) cores and photodegradable Newkome-type dendrons self-assemble into micron-sized ordered complexes with a face-centred-cubic superstructure. Interestingly, the magnetic properties of magnetoferritin NPs have been shown to be affected directly by the hierarchical organization. Additionally, the magnetoferritin-dendron assemblies efficiently disassembled by a short optical stimulus resulting in the release of free magnetoferritin NPs and restoration of the typical magnetic properties of magnetoferritin.

Protein	PDB ID_chain_Nb of residues (Resolution, Å)	Superimposed residues (Nb)	RMSD (Cα atoms)	Sequence identity (%)
Human L chain ferritin	2FFX_J_173 (1.90)	169	0.57	56
P. furiosus ferritin	2JD7_A_167 (2.80)	158	1.62	31
E. coli Bfr	2Y3Q_A_157 (1.55)	152	2.01	18
L. innocua Dps	1QGH_A_150 (2.35)	125	1.82	10

Table 1. Structure comparison between single subunits of proteins belonging to the ferritin family.

5. Conclusion

In conclusion, in this chapter we have shown that proteins belonging to the ferritin family, and in particular ferritins and Dps proteins, represent a rich and productive set of biomolecular templates for directed materials synthesis in the nano-scale. Indeed, a very wide range of inorganic materials has been successfully synthesized within these cage structures so far. Additionally, the protein shell has been manipulated both chemically and genetically to provide it with new functionalities. As a result, a large variety of materials have been produced, which can be applied in a number of fields as diverse as medicine, chemistry and electronics.

6. Acknowledgment

P.C. thanks the Associazione Italiana per la Ricerca sul Cancro (AIRC), grant agreement n° MFAG10545, for support of the research on biomedical applications of ferritin-based nanoparticles.

The crystal structure of HuHFt (PDB ID: 2CEI, resolution: 1.80 Å, chain A, 172 residues) has been structurally superimposed to the indicated protein subunits using the program "Protein structure comparison service Fold at European Bioinformatics Institute" (PDBeFold; http://www.ebi.ac.uk/msd-srv/ssm/; Krissinel & Henrick, 2004).

7. References

Aisen, P., Enns, C. & Wessling-Resnick, M. (2001) Chemistry and biology of eukaryotic iron metabolism. *Int. J. Biochem.Cell. Biol. Reviews,* 33, 10, 940-959

Alaleona, F., Franceschini, S., Ceci, P., Ilari, A. & Chiancone, E. (2010) Thermosynechococcus elongatus DpsA binds Zn(II) at a unique three histidine-containing ferroxidase center and utilizes O_2 as iron oxidant with very high efficiency, unlike thetypical Dps proteins. *FEBS J.,* 277, 4, 903-917

Almirón, M., Link, A.J., Furlong, D. & Kolter, R. (1992) A novel DNA-binding protein with regulatory and protective roles in starved Escherichia coli. *Genes Dev.,* 6, 12BA, 2646-2654

Andrews, S.C., Arosio, P., Bottke, W., Briat, J.F., von Darl, M., Harrison, P.M., Laulhère, J.P., Levi, S., Lobreaux, S. & Yewdall, S.J. (1992) Structure, function, and evolution of ferritins. *J. Inorg. Biochem.,* 47, 3-4, 161-174

Andrews, S.C., Robinson, A.K. & Rodríguez-Quiñones, F. (2003) Bacterial iron homeostasis. *FEMS Microbiol. Reviews,* 27, 2-3, 215-237

Andrews, S.K. (2010) The Ferritin-like superfamily: Evolution of the biological iron store man from a rubrerythrin-like ancestor. *Biochim. Biophys. Acta,* 1800, 8, 691-705

Babincová, M., Leszczynska, D., Sourivong, P. & Babinec, P. (2000) Selective treatment of neoplastic cells using ferritin-mediated electromagnetic hyperthermia. *Med. Hypotheses,* 54, 2, 177-179

Bellapadrona, G., Stefanini, S., Zamparelli, C., Theil, E.C. & Chiancone, E. (2009) Iron translocation into and out of Listeria innocua Dps and size distribution of the protein-enclosed nanomineral are modulated by the electrostatic gradient at the 3-fold "ferritin-like" pores. *J. Biol. Chem.,* 284, 28, 19101-19109

Bellapadrona, G., Ardini, M., Ceci, P., Stefanini, S. & Chiancone, E. (2010) Dps proteins prevent Fenton-mediated oxidative damage by trapping hydroxyl radicals within the protein shell. *Free Radic. Biol. Med.,* 48, 2,292-297

Bode, S.A., Minten, I.J., Nolte, R.J. & Cornelissen, J.J. (2011) Reactions inside nanoscale protein cages. *Nanoscale,* 3, 6, 2376-2389

Bou-Abdallah, F., Arosio, P., Levi, S., Janus-Chandler, C. & Chasteen, N.D. (2003) Defining metal ion inhibitor interactions with recombinant human H- and L-chain ferritins and site-directed variants: an isothermal titration calorimetry study. *J. Biol. Inorg. Chem.,* 8, 4, 489-497

Bou-Abdallah, F., Biasiotto, G., Arosio, P. & Chasteen, N.D. (2004) The putative "nucleation site" in human H-chain ferritin is not required for mineralization of the iron core. *Biochemistry,* 43, 14, 4332-4337

Bou-Abdallah, F., Zhao, G., Biasiotto, G., Poli, M., Arosio, P. & Chasteen, N.D. (2008) Facilitated diffusion of iron (II) and dioxygen substrates into human H-hain ferritin. A fluorescence and absorbance study employing the ferroxidase center substitution Y34W. *J. Am. Chem. Soc.,* 130, 52, 17801-17811

Braun, V. & Hantke, K. (2011) Recent insights into iron import by bacteria. *Curr. Opin. Chem.. Biol.,* 15, 328–334

Bunker, J., Lowry, T., Davis, G., Zhang, B., Brosnahan, D., Lindsay, S., Costen, R., Choi, S., Arosio, P. & Watt, G.D. (2005) Kinetic studies of iron deposition catalyzed by recombinant human liver heavy and light ferritins and Azotobacter vinelandii bacterioferritin using O_2 and H_2O_2 as oxidants. *Biophys. Chem.,* 114, 2-3, 235-44

Ceci, P., Ilari, A., Falvo, E. & Chiancone, E. (2003) The Dps protein of Agrobacterium tumefaciens does not bind to DNA but protects it toward oxidative cleavage: x-ray crystal structure, iron binding, and hydroxyl-radical scavenging properties. *J. Biol. Chem.*, 278, 22, 20319-20326

Ceci, P., Chiancone, E., Kasyutich, O., Bellapadrona, G., Castelli, L., Fittipaldi, M., Gatteschi, D., Innocenti, C. & Sangregorio, C. (2010) Synthesis of iron oxide nanoparticles in Listeria innocua Dps (DNA-binding protein from starved cells): a study with the wild-type protein and a catalytic centre mutant. *Chemistry*, 16, 2, 709-717

Chiancone, E. & Ceci, P. (2010) The multifaceted capacity of Dps proteins to combat bacterial stress conditions: Detoxification of iron and hydrogen peroxide and DNA binding. *Biochim. Biophys. Acta*, 1800, 8, 798-805

Chiaraluce, R., Consalvi, V., Cavallo, S., Ilari, A., Stefanini, S & Chiancone, E. (2000) The unusual dodecameric ferritin from Listeria innocua dissociates below pH 2.0. *Eur. J. Biochem.*, 267, 18, 5733-5741

Crichton, R.R. & Boelaert, J.R. (2009) Inorganic Biochemistry of Iron Metabolism from Molecular Mechanisms to Clinical Consequences, 3rd Ed., *John Wiley and Sons, Chichester*, 183–222

Crichton, R.R. & Declercq, J.P. (2010) X-ray structures of ferritins and related proteins. *Biochim. Biophys. Acta*, 1800, 8, 706-718

Crow, A., Lawson, T.L., Lewin, A., Moore, G.R. & Le Brun, N.E. (2009) Structural basis for iron mineralization by bacterioferritin. *J. Am. Chem. Soc.*, 131, 19, 6808-6813

Cuypers, M.G., Mitchell, E.P., Romão, C.V. & McSweeney, S.M. (2007) The crystal structure of the Dps2 from Deinococcus radiodurans reveals an unusual pore profile with anon-specific metal binding site. *J. Mol. Biol.*, 371, 3, 787-799

Doig, P., Austin, J.W. & Trust, T.J. (1993) The Helicobacter pylori 19.6-kilodalton protein is an iron-containing protein resembling ferritin. *J. Bacteriol.*, 175, 2, 557-560

Douglas, T., & Ripoll, D.R. (1998) Calculated electrostatic gradients in recombinant human H-chain ferritin. *Protein Sci.*, 7, 5, 1083-1091

Drysdale, J.W. (1976) Ferritin phenotypes: structure and metabolism. *Ciba Found Symp.*, 9, 51, 41-67

Franceschini, S., Ceci, P., Alaleona, F., Chiancone, E. & Ilari, A. (2006) Antioxidant Dps protein from the thermophilic cyanobacterium Thermosynechococcus elongatus. *FEBS J.*, 273, 21, 4913-4928

Frolow, F., Kalb, A.J. & Yariv, J. (1994) Structure of a unique two fold symmetric haem-binding site. *Nat. Struct. Biol.*, 1, 7, 453-60

Gálvez, N., Fernández, B., Sánchez, P., Cuesta, R., Ceolín, M., Clemente-León, M., Trasobares, S., López-Haro, M., Calvino, J.J., Stéphan, O. & Domínguez-Vera, J.M. (2008) Comparative structural and chemical studies of ferritin cores with gradual removal of their iron contents. *J. Am. Chem. Soc.*, 25, 130, 8062-8068

Geninatti, C.S., Crich, S., Bussolati, B., Tei, L., Grange, C., Esposito, G., Lanzardo, S., Camussi, G. & Aime, S. (2006) Magnetic resonance visualization of tumor angiogenesis by targeting neural cell adhesion molecules with the highly sensitive gadolinium-loaded apoferritin probe. *Cancer Res.*, 66, 18, 9196-9201

Gilmore, K., Idzerda, Y.U., Klem, M.T., Allen, M., Douglas, T. & Young, M. (2005) Surface contribution to the anisotropy energy of spherical magnetite particles. *J. Appl. Phys.*, 97, 10, 10B301 - 10B303

Granier, T., Langlois d'Estaintot, B., Gallois, B., Chevalier, J.M., Précigoux, G., Santambrogio, P. & Arosio, P. (2003) Structural description of the active sites of mouse L-chain ferritin at 1.2 A resolution. *J. Biol. Inorg. Chem.*, 8, 1-2, 105-111

Grant, R.A., Filman, D.J., Finkel, S.E., Kolter, R. & Hogle, J.M. (1998) The crystal structure of Dps, a ferritin homolog that binds and protects DNA. *Nat. Struct. Biol.*, 5, 4, 294-303

Grünberg, K., Wawer, C., Tebo, B.M., Shüler D. (2001) A large gene cluster encoding several magnetosome proteins is conserved in different species of magnetotactic bacteria. *Appl. Envir. Microbiol.* 67, 10, 4573-4582

Hainfeld, J.F. (1992) Uranium-loaded apoferritin with antibodies attached: molecular design for uranium neutron-capture therapy. *Proc. Natl. Acad. Sci.*, 89, 22, 11064-11068

Hikono, T., Matsumura, T., Miura, A., Uraoka, Y., Fuyuki, T., Takeguchi, M., Yoshii, S. & Yamashita, I. (2006) Electron confinement in a metal nanodot monolayer embedded in silicon dioxide produced using ferritin protein, *Appl. Phys. Lett.*, 88

Ichikawa, K., Uraoka, Y., Punchaipetch, P., Yano, H., Hatayama, T., Fuyuki, T. & Yamashita, I. (2007) Low-temperature polycrystalline silicon thin film transistor flash memory with ferritin. *Jpn. J. Appl. Phys.*, 46, 34, L804-L806

Ilari, A., Stefanini, S., Chiancone, E. & Tsernoglou, D. (2000) The dodecameric ferritin from Listeria innocua contains a novel intersubunit iron-binding site. *Nat. Struct. Biol.*, 7, 1, 38-43

Ilari, A., Ceci, P., Ferrari, D., Rossi, G.L. & Chiancone E. (2002) Iron incorporation into Escherichia coli Dps gives rise to a ferritin-like microcrystalline core. *J. Biol. Chem.*, 277, 40, 37619-37623

Joh, D.Y., Herman, L.H., Ju, S.Y., Kinder, J., Segal, M.A., Johnson, J.N., Chan, G.K. & Park, J. (2011) On-chip Rayleigh imaging and spectroscopy of carbon nanotubes. *Nano Lett.*, 11, 1, 1-7

Johnson, E., Cascio, D., Sawaya, M.R., Gingery, M. & Schröder, I. (2005) Crystal structures of a tetrahedral open pore ferritin from the hyperthermophilic archaeon Archaeoglobus fulgidus. *Structure*, 13, 4, 637-648

Kim, I., Hosein, H.A., Strongin, D.S. & Douglas, T. (2002) Photochemical reactivity of ferritin for Cr (VI) reduction. *Chemistry of Materials*, 14, 4874-4879

Klem, M.T., Resnick, D.A., Gilmore, K., Young M., Idzerda, Y.U. & Douglas, T. (2007) Synthetic control over magnetic moment and exchange bias in all-oxide materials encapsulated within a spherical protein cage. *J. Am. Chem. Soc.*, 129, 1, 197–201

Kauko, A., Pulliainen, A.T., Haataja, S., Meyer-Klaucke, W., Finne, J. & Papageorgiou, A.C. (2006) Iron incorporation in Streptococcus suis Dps-like peroxide resistance protein Dpr requires mobility in the ferroxidase center and leads to the formation of a ferrihydrite-like core. *J. Mol. Biol.*, 364, 1, 97-109

Kasyutich, O., Ilari, A., Fiorillo, A., Tatchev, D., Hoell, A. & Ceci, P. (2010) Silver ion incorporation and nanoparticle formation inside the cavity of Pyrococcus furiosus ferritin: structural and size-distribution analyses. *J. Am. Chem. Soc.*, 132, 10, 3621-3627

Kostiainen, M.A., Ceci, P., Fornara, M., Hiekkataipale, P., Kasyutich, O., Nolte, R.J.,. Cornelissen, J.J, Desautels, R.D. & van Lierop, J. (2011) Hierarchical Self-Assembly and Optical Disassembly for Controlled Switching of Magnetoferritin Nanoparticle Magnetism. *ACS Nano, in press*

Krissinel, E., Henrick, K. (2004) Secondary-structure matching (SSM), a new tool for fast protein structure alignment in three dimensions. *Acta Crystallogr D Biol Crystallogr.* 60, 2256-2268

Lawson, D.M., Artymiuk, P.J., Yewdall, S.J., Smith, J.M., Livingstone, J.C., Treffry, A., Luzzago, A., Levi, S., Arosio, P., Cesareni, G. & et al. (1991) Solving the structure of human H-ferritin by genetically engineering intermolecular crystal contacts. *Nature,* 349, 541–544

Lawson, D.M., Treffry, A., Artymiuk, P.J., Harrison, P.M., Yewdall, S.J., Luzzago, A., Cesareni, G., Levi, S. & Arosio, P. (1989) Identification of the ferroxidase centre in ferritin. *FEBS Lett.,* 254, 1-2, 207-210

Le Brun, N.E., Crow, A., Murphy, M.E., Mauk, A.G. & Moore, G.R. (2010) Iron core mineralisation in prokaryotic ferritins. *Biochim Biophys Acta,* 1800, 8, 732-744.

Li, W., Xu, L.H., Forssell, C., Sullivan, J.L. & Yuan, X.M. (2008) Overexpression of transferring receptor and ferritin related to clinical symptoms and destabilization of human carotid plaques. *Exp. Biol. Med.,* 233, 7, 818-826

Lin, X., Xie, J., Niu, G., Zhang, F., Gao, H., Yang, M., Quan, Q., Aronova, M.A., Zhang, G., Lee, S., Leapman, R. & Chen, X. (2011) Chimeric ferritin nanocages for multiple function loading and multimodal imaging. *Nano Lett.,* 11, 2, 814-819

Liu, H-l., Zhou, H.N., Xing, W.M., Zhao, J.F., Li, S.X., Huang, J.F. & Bi, R.C. (2004) 2.6 A resolution crystal structure of the bacterioferritin from Azotobacter vinelandii. *FEBS Lett.,* 573, 1-3, 93-98

Liu, X. & Theil, E.C. (2005) Ferritins: dynamic management of biological iron and oxygen chemistry. *Acc. Chem. Res.,* 38, 3, 167-175

Lewin, A., Moore, G.R. & Le Brun, N.E. (2005) Formation of protein-coated iron minerals. *Dalton Trans.,* 22, 3597–3610

Macara, I.G., Hoy, T.G. & Harrison, P.M. (1973) The formation of ferritin from apoferritin. Inhibition and metal ion-binding studies. *Biochem. J.,* 135, 4, 785-789.

Mann, S., Spark, N.H.C., Board, R.G. (1990) Magnetotactic Bacteria: microbiology, biomineralization, palaeomagnetism and biotechnology. *Adv. Microbiol. Physiol.,* 31, 125-181.

Massover, W.H. & Cowley, J.M. (1973) The ultrastructure of ferritin macromolecules. The lattice structure of the core crystallites. *Proc. Natl. Acad. Sci U.S.,* 70, 12, 3847-3851

McCord, J.M. & Fridovich, I. (1988) Superoxide dismutase: the first twenty years (1968-1988). *Free Radic. Biol. Med.,* 5, 5-6,363-369

Miura, A., Hkono, T., Matsumura, T., Yano, H., Hatayama, T., Uraoka, Y., Fuyuki, T., Yoshii, S. & Yamashita I. (2006) Floating nanodot gate memory devices based on biomineralized inorganic nanodot array as a storage node. *Jpn. J. Appl. Phys.,* 45, L1–L3

Nordlund, P. & Eklund, H. (1995) Di-iron-carboxylate proteins. *Curr. Opin. Struct. Biol. Reviews,* 5, 6, 758-766

Prastaro, A., Ceci, P., Chiancone, E., Boffi, A., Cirilli, R., Colone, M., Fabrizi, G., Stringaro, A. & Cacchi, S. (2009) Suzuki-Miyaura cross-coupling catalyzed by protein-stabilized palladium nanoparticles under aerobic conditions in water: application to a one-pot chemoenzymatic enantioselective synthesis of chiral biaryl alcohols. *Green Chemistry,* 11, 1929-1932

Prastaro, A., Ceci, P., Chiancone, E., Boffi, A., Fabrizi, G.& Cacchi, S. (2010) Homocoupling of arylboronic acids and potassium aryltrifluoroborates catalyzed by protein-stabilized palladium nanoparticles under air in water. *Tetrahedron Letters,* 51, 18, 2550-2552

Ren, B., Tibbelin, G., Kajino, T., Asami, O. & Ladenstein, R. (2003) The multi-layered structure of Dps with a novel di-nuclear ferroxidase center. *J. Mol. Biol.,* 329, 3, 467-477

Rohrer, J.S., Islam, Q.T., Watt, G.D., Sayers, D.E. & Theil, E.C. (1990) Iron environment in ferritin with large amounts of phosphate, from Azotobacter vinelandii and horse spleen, analyzed using extended X-ray absorption fine-structure (EXAFS). *Biochemistry*, 29, 1, 259–264

Romåo, C.V., Regalla, M., Xavier, A.V., Teixeira, M., Liu, M.Y. & Le Gall, J.A. (2000) Bacterioferritin from the strict anaerobe Desulfovibrio desulfuricans ATCC 27774. *Biochemistry*, 39, 23, 6841-6849

Romão, C.V., Mitchell, E.P. & McSweeney, S. (2006) The crystal structure of Deinococcus radiodurans Dps protein (DR2263) reveals the presence of a novel metal centre in the N terminus. *J. Biol. Inorg. Chem.*, 11, 7, 891-902

Sánchez, P., Valero, E., Gálvez, N., Domínguez-Vera, J.M., Marinone, M., Poletti, G., Corti, M. & Lascialfari, A. (2009) MRI relaxation properties of water-soluble apoferritin-encapsulated gadolinium oxide hydroxide nanoparticles. *Dalton Trans.*, 7, 5, 800-804.

Santambrogio, P., Levi, S., Cozzi, A., Corsi, B. & Arosio, P. (1996) Evidence that the specificity of iron incorporation into homopolymers of human ferritin L- and H-chains is conferred by the nucleation and ferroxidase centres. *Biochem. J.*, 314, Pt 1, 139-144.

Stefanini, S., Desideri, A., Vecchini, P., Drakenberg, T. & Chiancone, E. (1989) Identification of the iron entry channels in apoferritin. Chemical modification and spectroscopic studies. *Biochemistry*, 28, 1, 378-382

Stefanini, S., Cavallo, S., Montanini, B. & Chiancone E. (1999) Incorporation of iron by the unusual dodecameric ferritin from Listeria innocua. *Biochem. J.*, 338, Pt 1, 71-75

Stillman, T.J., Hempstead, P.D., Artymiuk, P.J., Andrews, S.C., Hudson, A.J., Treffry, A., Guest, J.R. & Harrison, P.M. (2001) The high-resolution X-ray crystallographic structure of the ferritin (EcFtnA) of Escherichia coli; comparison with human H ferritin (HuHF) and the structures of the Fe(3+) and Zn(2+) derivatives. *J. Mol. Biol.*, 307, 2, 587-603

Suzuki, M., Abe, M., Ueno, T., Abe, S., Goto, T., Toda, Y., Akita, T., Yamada, Y. & Watanabe,Y. (2009) Preparation and catalytic reaction of Au/Pd bimetallic nanoparticles in apo-ferritin. *Chem Commun. (Camb)*, 32, 4871-4873

Tatur, J., Hagen, W.R. & Matias, P.M. (2007) Crystal structure of the ferritin from the hyperthermophilic archaeal anaerobe Pyrococcus furiosus. *J. Biol. Inorg. Chem.*, 12, 5, 615-630

Terashima, M., Uchida, M., Kosuge, H., Tsao, P.S., Young, M.J., Conolly, S.M., Douglas, T. & McConnell, M.V. (2011) Human ferritin cages for imaging vascular macrophages. *Biomaterials*, 32, 5, 1430-1437

Toussaint, L., Bertrand, L., Hue, L., Crichton, R.R. & Declercq, J.P. (2007) High-resolution X-ray structures of human apoferritin H-chain mutants correlated with their activity and metal-binding sites. *J. Mol. Biol.*, 365, 2, 440-452

Treffry, A., Bauminger, E.R., Hechel, D., Hodson, N.W., Nowik, I., Yewdall, S.J. & Harrison, P.M. (1993) Defining the roles of the threefold channels in iron uptake, iron oxidation and iron-core formation in ferritin: a study aided by site-directed mutagenesis. *Biochem. J.*, 296, Pt 3, 721-728

Trikha, J., Theil, E.C. & Allewell, N.M. (1995) High resolution crystal structures of amphibian red-cell L ferritin: potential roles for structural plasticity and solvation in function. *J. Mol. Biol.*, 248, 5, 949-967

Uchida M., Flenniken, M.L., Allen, M., Willits, D.A., Crowley, B.E., Brumfield, S., Willis, A.F., Jackiw, L., Jutila, M., Young, M.J. & Douglas, T. (2006) Targeting of cancer cells with ferrimagnetic ferritin cage nanoparticles. *J. Am .Chem. Soc.*, 128, 51, 16626-16633

Uchida, M., Terashima, M., Cunningham, C.H., Suzuki, Y., Willits, D.A., Willis, A.F., Yang, P.C., Tsao, P.S., McConnell, M.V., Young, M.J. & Douglas, T. (2008) A human ferritin iron oxide nano-composite magnetic resonance contrast agent. *Magn. Reson. Med.*, 60, 5, 1073-1081

Uchida, M., Kang, S., Reichhardt, C., Harlen, K. & Douglas, T. (2010) The ferritin superfamily: Supramolecular templates for materials synthesis. *Biochim. Biophys. Acta*, 1800, 8, 834-845

Ueno, T., Suzuki, M., Goto, T., Matsumoto, T., Nagayama, K. & Watanabe, Y. (2004) Size-selective olefin hydrogenation by a Pd nanocluster provided in an apo-ferritin cage. *Angew. Chem. Int. Ed.*, 43, 19, 2527-2530

Ueno, T. Abe, M., Hirata, K., Abe, S., Suzuki, M., Shimizu, N., Yamamoto, M., Takata, M. & Watanabe, Y. (2009) Process of accumulation of metal ions on the interior surface of apo-ferritin: crystal structures of a series of apo-ferritins containing variable quantities of Pd (II) ions *J. Am. Chem. Soc.*, 131, 14, 5094-5100

Wardeska, J.G., Viglione, B. & Chasteen, N.D. (1986) Metal ion complexes of apoferritin. Evidence for initial binding in the hydrophilic channels. *J. Biol. Chem.*, 261, 15, 6677-6683

Watt, G.D., Jacobs, D. & Frankel, R.B. (1988) Redox reactivity of bacterial and mammalian ferritin: is reductant entry into the ferritin interior a necessary step for iron release? *Proc. Natl. Acad. Sci U S A*, 85, 20, 7457-7461

Xing, R., Wang, X., Zhang, C., Zhang, Y., Wang, Q., Yang, Z. & Guo, Z. (2009) Characterization and cellular uptake of platinum anticancer drugs encapsulated in apoferritin. *J. Inorg. Biochem.*, 103, 7, 1039-1044

Xu, B. & Chasteen, N.D. (1991) Iron oxidation chemistry in ferritin. Increasing Fe/O_2 stoichiometry during core formation. *J. Biol. Chem.*, 266, 30, 19965-19970

Yang, X. & Chasteen, N.D. (1999) Ferroxidase activity of ferritin: effects of pH, buffer and Fe (II) and Fe(III) concentrations on Fe(II) autoxidation and ferroxidation. *Biochem. J.*, 338, Pt 3, 615-618

Yang, Z., Wang, X., Diao, H., Zhang, J., Li, H., Sun, H. & Guo, Z. (2007) Encapsulation of platinum anticancer drugs by apoferritin. *Chem. Commun. (Camb.)*, 7, 33, 3453-3455

Yamamoto, Y., Poole, L.B., Hantgan, R.R. & Kamio, Y. (2002) An iron-binding protein, Dpr, from Streptococcus mutans prevents iron-dependent hydroxyl radical formation in vitro. *J. Bacteriol.*, 184, 11, 2931-2939

Yamada, K., Yoshii, S., Kumagai, S., Miura, A., Uraoka, Y., Fuyuki, T. & Yamashita, I. (2007) Effects of dot density and dot size on charge injection characteristics in nanodot array produced by protein supramolecules, *Jpn. J. Appl. Phys.*, 46, 7549–7553

Yamashita, I. (2008) Biosupramolecules for nano-devices: biomineralization of nanoparticles and their applications. *J. Mater. Chem.*, 18, 3813–3820

Zanotti, G., Papinutto, E., Dundon, W., Battistutta, R., Seveso, M., Giudice, G., Rappuoli, R. & Montecucco, C. (2002) Structure of the neutrophil-activating protein from Helicobacter pylori. *J. Mol. Biol.*, 323, 1, 125-130

Pymol (Delano Scientific LLC, San Carlos, LA; http://www.pymol.org)

7

Biofilm and Microbial Applications in Biomineralized Concrete

Navdeep Kaur Dhami, Sudhakara M. Reddy and Abhijit Mukherjee
Thapar University, Patiala
India

1. Introduction

Biomineralization is defined as a biologically induced process in which an organism creates a local micro-environment with conditions that allow optimal extracellular chemical precipitation of mineral phases (Hamilton, 2003). The synthesis of these minerals by prokaryotes is broadly classified into two classes: Biologically controlled mineralization (BCM) and Biologically induced mineralization (BIM) (Lowenstam, 1981; Lowenstam & Weiner, 1989). In the case of biologically controlled mineralization, minerals are directly synthesized at a specific location within or on the cell and only under certain conditions. In most cases, BCM happens intracellularly, where lipids, proteins, polysaccharides, etc. make a stable matrix for cations to condense and minerals to grow in a constrained space. Minerals that form by biologically induced mineralization processes generally nucleate and grow extracellularly as a result of metabolic activity of the organism and subsequent chemical reactions involving metabolic byproducts. Bacterial surfaces such as cell walls or polymeric materials (exopolymers) exuded by bacteria includes slimes, sheaths, or biofilms, or even dormant spores, can act as important sites for the adsorption of ions and mineral nucleation and growth (Beveridge, 1989; Konhauser, 1998; Banfield & Zhang, 2001; Bäuerlein, 2003).

Bacterially induced and mediated mineralization is a research subject widely studied in the past decades (Banfield & Hamers, 1997; Douglas & Beveridge, 1998; Ehrlich, 2002). Due to its numerous consequences, bacterially induced precipitation of calcium carbonate, so-called carbonatogenesis (Rodriguez-Navarro et al., 2003), has attracted much attention from both basic and applied points of view. It has implications for: (1) atmospheric CO_2 fixation through carbonate sediment formation and lithification (Krumbein, 1979; Monger et al., 1991; Chafetz & Buczynski, 1992; Folk, 1993) and dolomite precipitation (Vasconcelos et al., 1995) (2) solid-phase capture of inorganic contaminants (Warren et al., 2001) (3) pathological formation of mineral concretions, such as gallstones and kidney stones in humans (Keefe, 1976) (4) the possibility of understanding extraterrestrial biological processes such as Martian carbonate production by bacteria (McKay et al., 1996; Thomas-Keprta et al., 1998).

Bacterially induced mineralization has recently emerged as a method for protecting and consolidating decayed construction materials. In nature, building and remediation of structures with local materials occurs without any requirement of extreme energy usage. Calcification and polymerization occur at ambient conditions as can be seen from the sustainability of ant hills and coral reefs. This occurs through the application of microorganisms which deposit carbonates (as part of their basic metabolic activities), one of the most well known minerals. These deposits (commonly called as calcium carbonate crystals/ calcite crystals/ microbial concrete) act as binders between loose substrate particles and reduce the pores inside the substrate particles. The use of bacteria for remediating building materials seems like a new idea, but this conservation method mimics what nature has been doing for eons, since many carbonate rocks have been cemented by calcium carbonate precipitation from microbes. The technology of using microbes for calcium carbonate deposition or microbial concrete, called as Microbially induced calcium carbonate precipitation (MICCP) can be used for solving various durability issues of construction materials. Microorganisms are abundant in nature, which paves the way for massive production of bacterial calcium carbonate crystals/ calcite/ concrete. As the microorganisms can penetrate and reproduce themselves in soil or any such environments, there is no need to disturb the ground or environment unlike that of cement. This technology also offers the benefit of being novel and eco- friendly.

Undoubtedly, broad a range of products are available in the market for protection of concrete surfaces (Basheer et al., 1997; Ibrahim et al., 1997; Basheer et al., 2006). Several of these products are organic coatings consisting of volatile organic compounds. The air polluting effect of these compounds during manufacturing and coating has led to the development of new formulations such as inorganic coating materials. Traditional inorganic coatings consist of calcium-silicate compounds, which exhibit a composition similar to cement (Moon et al., 2007). Surface treatments with water repellants like epoxy injections, with pore blockers and various synthetic agents like silanes or siloxanes are also available today, but with a number of disadvantages like degradation with time, need for constant maintenance and environment pollution (De Muynck et al., 2006).

In the case of carbonate stones like limestone, dolostone, and marble, progressive dissolution of the mineral matrix as a consequence of weathering leads to an increase of the porosity, and as a result, a decrease of the mechanical features (Tiano et al., 1999). In order to decrease the susceptibility to decay, many conservation treatments have been applied with the aim of modifying some of the stone characteristics. Water repellents have been applied to protect stone from the ingress of water and other weathering agents. The use of stone consolidants aim at re-establishing the cohesion between separated grains of deteriorated stone. Due to problems related to incompatibility with the stone, both water repellents and consolidants have often been reported to accelerate stone decay (Clifton & Frohnsdorff, 1982; Delgado Rodriguez, 2001; Moropoulou et al., 2003).

Another issue related with conventional building materials is the high production of green house gases and high energy consumed during production of these materials. The emission of green house gases during manufacturing processes of building materials is contributing a detrimental amount to global warming. Along with this, high construction cost of building materials is another issue that needs to be dealt with.

The above mentioned drawbacks of conventional treatments have invited the usage of novel, eco- friendly, self healing and energy efficient technology where microbes are used for remediation of building materials and enhancement in the durability characteristics. This technology may bring new approaches in the construction industry.

2. Production of microbial concrete/calcite

Bacteria from various natural habitats have frequently been reported to be able to precipitate calcium carbonate both in natural and in laboratory conditions (Krumbein, 1979; Rodriguez et al., 2003). Different types of bacteria, as well as abiotic factors (salinity and composition of the medium) seem to contribute in a variety of ways to calcium carbonate precipitation in a wide range of different environments (Knorre & Krumbein, 2000; Rivadeneyra et al., 2004). Calcium carbonate precipitation is a straight forward chemical process governed mainly by four key factors: (1) the calcium concentration, (2) the concentration of dissolved inorganic carbon (DIC), (3) the pH and (4) the availability of nucleation sites (Hammes & Verstraete, 2002). $CaCO_3$ precipitation requires sufficient calcium and carbonate ions so that the ion activity product (IAP) exceeds the solubility constant (K_{so}) (Eqs. (1) and (2)). From the comparison of the IAP with the K_{so} the saturation state (Ω) of the system can be defined; if $\Omega > 1$ the system is oversaturated and precipitation is likely (Morse, 1983):

$$Ca^{2+} + CO_3{}^{2-} \longleftrightarrow CaCO_3 \tag{1}$$

$$\Omega = a(Ca^{2+})a(CO_3{}^{2-}) / K_{so} \text{ with } K_{so\ calcite, 25^\circ} = 4.8 \times 10^{-9} \tag{2}$$

The concentration of carbonate ions is related to the concentration of DIC and the pH of a given aquatic system. In addition, the concentration of DIC depends on several environmental parameters such as tempeqrature and the partial pressure of carbon dioxide (for systems exposed to the atmosphere). The equilibrium reactions and constants governing the dissolution of CO_2 in aqueous media (25°C and 1 atm) are given in Eqs. (3)–(6) (Stumm & Morgan, 1981):

$$CO_{2(g)} \longleftrightarrow CO_{2(aq.)} \ (pK_H = 1.468) \tag{3}$$

$$CO_{2(aq.)} + H_2O \longleftrightarrow H_2CO_3 * (pK = 2.84) \tag{4}$$

$$H_2CO_3 * \longleftrightarrow H^+ + HCO_3{}^- \ (pK1 = 6.352) \tag{5}$$

$$HCO_3{}^- \longleftrightarrow CO_3{}^{2-} + H^+ \ (pK2 = 10.329) \tag{6}$$

With $$H_2CO_3 * = CO_{2(aq.)} + H_2CO_3$$

Microorganisms can influence precipitation by altering almost any of the precipitation parameters described above, either separately or in various combinations with one another (Hammes & Verstraete, 2002).

Different pathways appear to be involved in calcium carbonate precipitation. The first pathway involves the sulphur cycle, in particular sulphate reduction, which is carried out by sulphate reducing bacteria under anoxic conditions. A second pathway involves the nitrogen cycle, and more specifically, (1) the oxidative deamination of amino acids in aerobiosis, (2) the reduction of nitrate in anaerobiosis or microaerophily and (3) the degradation of urea or uric acid in aerobiosis (by ureolytic bacteria). Another microbial process that leads to an increase of both pH and the concentration of dissolved inorganic carbon is the utilization of organic acids (Braissant et al., 2002), a process which has been commonly used in microbial carbonate precipitation experiments. The precipitation pathways described in the aforementioned are generally found in nature which accounts for the common occurrence of microbial carbonate precipitation (MCP) and validates the statement by Boquet et al (1973) that under suitable conditions, most bacteria are capable of inducing carbonate precipitation. Due to the simplicity, the most commonly studied system of applied MICCP is urea hydrolysis via the enzyme urease in a calcium rich environment. Urease catalyzes the hydrolysis of urea to CO_2 and ammonia, resulting in an increase of pH and carbonate concentration in the bacterial environment. During microbial urease activity, 1 mol of urea is hydrolyzed intracellularly to 1 mol of ammonia and 1 mol of carbonate (Eq.7), which spontaneously hydrolyzes to form additional 1 mol of ammonia and carbonic acid (Eq.8) as follows:

$$CO(NH_2)_2 + H_2O \xrightarrow{bacteria} NH_2COOH + NH_3 \tag{7}$$

$$NH_2COOH + H_2O \longrightarrow NH_3 + H_2CO_3 \tag{8}$$

These products equilibrate in water to form bicarbonate, 1 mol of ammonium and hydroxide ions which give rise to pH increase

$$H_2CO_3 \longrightarrow 2H^+ + 2CO_3^{2-} \tag{9}$$

$$NH_3 + H_2O \longrightarrow NH^{4-} + OH^- \tag{10}$$

$$Ca^{2+} + CO_3^{2-} \longrightarrow CaCO_3 (K_{SP} = 3.8 \times 10^{-9}) \tag{11}$$

K_{SP} is the solubility product in Eq.11.

Hammes & Verstraete (2002) investigated the series of events occurring during ureolytic calcification emphasizing the importance of pH and calcium metabolism during the process (Fig.1). The primary role of bacteria has been ascribed to their ability to create an alkaline environment through various physiological activities.

Bacterial surfaces play an important role in calcium precipitation (Fortin et al., 1997). Due to the presence of several negatively charged groups, at a neutral pH, positively charged metal ions could be bound on bacterial surfaces, favouring heterogenous nucleation (Douglas, 1998; Bauerlein, 2003). Commonly, carbonate precipitates develop on the external surface of bacterial cells by successive stratification (Pentecost & Bauld, 1988; Castanier et al., 1999) and bacteria can be embedded in growing carbonate crystals (Rivadeneyra e al., 1998; Castanier et al., 1999).

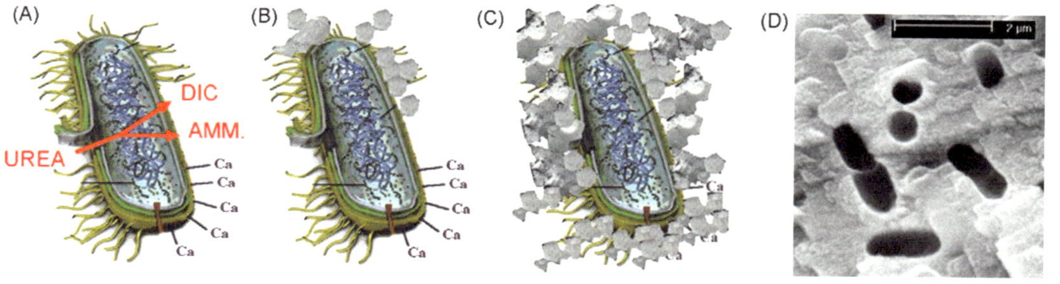

Fig. 1. Simplified representation of the events occurring during the microbially induced carbonate precipitation. Calcium ions in the solution are attracted to the bacterial cell wall due to the negative charge of the latter. Upon addition of urea to the bacteria, dissolved inorganic carbon (DIC) and ammonium (AMM) are released in the microenvironment of the bacteria (A). In the presence of calcium ions, this can result in a local supersaturation and hence heterogeneous precipitation of calcium carbonate on the bacterial cell wall (B). After a while, the whole cell becomes encapsulated (C), limiting nutrient transfer, resulting in cell death. Image (D) shows the imprints of bacterial cells involved in carbonate precipitation (Source: Hammes & Verstraete., 2002).

Possible biochemical reactions in urea-$CaCl_2$ medium to precipitate $CaCO_3$ at the cell surface can be summarized as follows:

$$Ca^{2+} + Cell \longrightarrow Cell - Ca^{2+} \tag{12}$$

$$Cl^{-} + HCO^{3-} + NH_3 \longrightarrow NH_4Cl + CO_3^{\,2-} \tag{13}$$

$$Cell - Ca^{2+} + CO_3^{\,2-} \longrightarrow Cell - CaCO_3 \tag{14}$$

The actual role of the bacterial precipitation remains, however, a matter of debate. Some authors believe this precipitation to be an unwanted and accidental by-product of the metabolism (Knorre & Krumbein, 2000) while others think that it is a specific process with ecological benefits for the precipitating organisms (Ehrlich, 1996; Mc Connaughey & Whelan, 1997).

However, a number of applications involving MICCP have been attempted as in the removal of heavy metals (Warren et al., 2001) and radionucleotides (Fujita et al., 2004), removal of calcium from wastewater (Hammes et al., 2003) and biodegradation of pollutants (Simon et al., 2004; Chaturvedi et al., 2006). Another series of applications aims at modifying the properties of soil, i.e. for the enhancement of oil recovery from oil reservoirs (Nemati & Voordouw, 2003; Nemati et al., 2005), plugging (Ferris & Stehmeier, 1992) and strengthening of sand columns (DeJong et al., 2006; Whiffin et al., 2007). Moreover, microbially induced precipitation has been investigated for its potential to improve the durability of construction materials such as cementitious materials, sand, bricks and limestone.

This technology has been investigated for its potential in consolidation and restoration of various building materials by many research groups on cement mortar cubes (Achal et al.,

2011b), sand consolidation and limestone monument repair (Stocks - Fischer et al., 1999; Bachmeier et al., 2002; Dick et al., 2006; Achal et al., 2009b), reduction of water and chloride ion permeability in concrete (Achal et al., 2011a), filling of pores and cracks in concrete (Bang et al., 2001; Ramakrishnan 2007; De Muynck et al., 2008a,b) and enhanced strength of red bricks (Sarda et al., 2009).

3. Applications of microbial concrete

The use of microbial concrete in Bio Geo Civil Engineering has become increasingly popular. From enhancement in durability of cementious materials to improvement in sand properties, from repair of limestone monuments, sealing of concrete cracks to highly durable bricks, microbial concrete has been successful in one and all. This new technology can provide ways for low cost and durable roads, high strength buildings with more bearing capacity, long lasting river banks, erosion prevention of loose sands and low cost durable housing. The next section will illustrate detailed analysis of role of microbial concrete in affecting the durability of building structures.

3.1 Microbial concrete in cementitious materials

Concrete is a strong and relatively cheap construction material used world wide. It is estimated that cement production alone contributes 7% to global anthropogenic CO_2 emissions, what is particularly due to the sintering of limestone and clay at a temperature of 1500°C, as during this process calcium carbonate is converted into calcium oxide while releasing CO_2 (Worrell et al., 2001). There is a great concern and emphasis in reducing the greenhouse gases emission into the atmosphere in order to control adverse environmental impacts.

Another aspect of concrete is its liability to cracking, a phenomenon that hampers the material's structural integrity and durability. It is generally accepted that the durability of concrete is related to the characteristics of its pore structure (Khan, 2003). Degradation mechanisms of concrete often depend on the way potentially aggressive substances can penetrate into the concrete, possibly causing damage. The permeability of concrete depends on the porosity and connectivity of the pores. The more open the pore structure of the concrete, the more vulnerable the material is to degradation mechanisms caused by penetrating substances. The deterioration of concrete structures usually involves movement of aggressive gases and/or liquids from the surrounding environment into the concrete, followed by physical and/or chemical reactions within its internal structure, possibly leading to irreversible damage (Claisse et al., 1997). The quality of concrete structures depends majorly on three parameters: compressive strength, permeability & corrosion. Crack problems in concrete are mostly dealt by manual inspections and repair by impregnation of cracks with epoxy based fillers, latex binding agents like acrylic, polyvinyl acetate, butadiene styrene etc. But there are many disadvantageous aspects of traditional repair systems such as different thermal expansion coefficient compared to concrete, weak bonding, environmental and health hazards along with being costly. So many researchers investigated the application of bacterial calcite in enhancing the durability of cementitious buildings and restoration of structures. Overview of different applications of microbial concrete in cementitious materials is given in table 1.

Application	Organism	Reference
Cement mortar and Concrete	*Bacillus cereus* *Bacillus sp. CT-5 Bacillus pasteurii* *Shewanella* *Sporosarcina pasteurii*	Le Metayer- Leverel et al (1999) Achal et al., 2011b Ramachandran et al (2001) Ghosh et al (2005) Achal et al (2011a)
Remediation of cracks in concrete	*Sporosarcina pasteurii* *Bacillus pasteurii* *Bacillus pasteurii* *Bacillus sphaericus* *Bacillus sphaericus*	Bang et al (2001) Ramachandran et al (2001) Ramakrishnan (2007) De Belie et al (2008) De Muynck et al (2008a, b)
Self Healing	*Bacillus pseudifirmus* *Bacillus cohnii*	Jonkers et al (2007)

Table 1. Overview of different applications where microbial concrete is used as biocement in cementitious materials.

3.1.1 Improvement in compressive strength of concrete

Compressive strength is one of the most important characteristic of concrete durability. It is considered as an index to assess the quality of concrete. More is the compressive strength, more is the durability of concrete specimen. Compressive strength test results are used to determine that the concrete mixture as delivered meets the requirements of the job specification. So, the effect of microbial concrete on compressive strength of concrete and mortar was studied and it was observed that significant enhancement in the strength of concrete and mortar can be seen upon application of bacteria.

The applicability of microbial concrete to affect the compressive strength of mortar and concrete was done by several studies (Bang et al., 2001; Ramachandran et al., 2001; Ghosh et al., 2005; De Muynck et al., 2008a,b; Jonkers et al., 2010; Achal et al., 2011b) where different microorganisms have been applied in the concrete mixture. Ramchandran et al (2001) observed the increase in compressive strength of cement mortar cubes at 7 and 28 days by using various concentrations of *Bacillus pasteurii*. They found that increase of strength resulted from the presence of adequate amount of organic substances in the matrix due to microbial biomass. Ghosh et al (2005) studied the positive potential of *Shewanella* on compressive strength of mortar specimens and found that the greatest improvement was at cell concentration of 10^5 cells/ ml for 3, 7, 14 and 28 days interval. They reported an increase of 17% and 25% after 7 and 28 days. But no noticeable increase was recorded in case of specimens treated with *Escherichia coli* (low urease producing). This concluded that choice of microorganism plays the prime role in improvement of strength characteristics. Jonkers & Schlangen (2007) studied the addition of bacterial spores ($10^8/cm^3$) of *Bacillus pseudofirmus* and *Bacillus cohnii* to concrete specimens and reported an increase of 10% in the compressive strength. Achal et al (2009a) treated mortar cubes with *Sporosarcina pasteurii* and observed 17% improvement in compressive strength. Another research group by Park et al (2010) observed 22% increase in the strength of mortar cubes treated by *Arthrobacter crystallopoietes* which was the higher compared to *Sporosarcina soli, Bacillus massiliensis and Lysinibacillus fusiformis* used along with. Upon addition of *Bacillus* sp. CT-5 to cement mortar specimens, 36% increase in compressive strength was reported (Achal et al., 2011b). Significant increase in compressive strength was observed at the age of 28 days as compared to earlier days. Microbial calcite might have precipitated on the surface of cells and eventually within the pores leading to their plugging which further lead to stoppage of flow of oxygen and nutrients to the cells. The cells either die or turn into endospores and act as organic fiber which enhances the compressive strength of mortar cubes (Ramachandran et al., 2001).

Further studies to confirm the role of microbial concrete in cement mortar specimens and concrete specimens were done by scanning electron microscope (Fig. 2), energy dispersive X-Ray spectrum (EDX) and X-ray diffraction (XRD) analysis of calcite crystals precipitated on concrete surfaces. The results confirmed the presence of calcite in the newly formed layer on the surface of concrete specimens (Ramachandran et al., 2001; Ghosh et al., 2005; De Muynck et al., 2008a; Achal et al., 2009a, 2011). Rod shaped bacteria were found embedded in calcite crystals which proved that bacteria act as the source of nucleation.

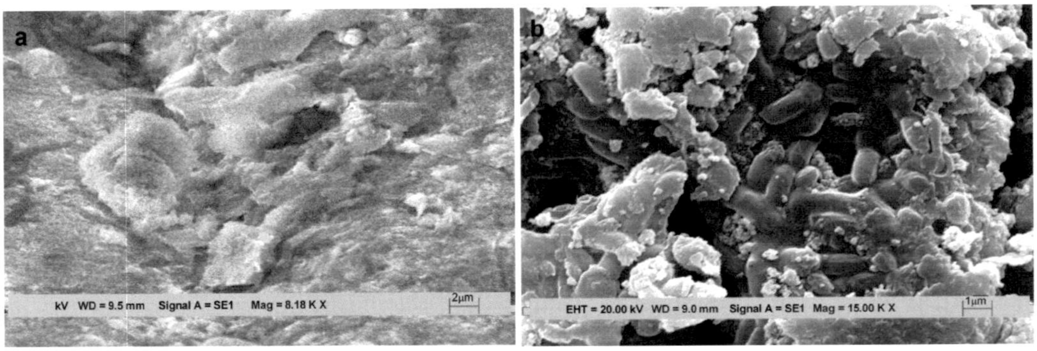

Fig. 2. Scanning electron micrographs of cement mortar specimens: (a) matrix of cement mortar prepared without bacteria (b) showing dense calcite precipitation as calcite crystals with rod-shaped impressions housed by *Bacillus sp. CT-5* (Achal et al., 2011b).

3.1.2 Reduction in permeability

Permeability of concrete is another important characteristic of concrete that affects its durability. Concrete with low permeability has been reported to last longer (Nolan et al., 1995). Permeation is required for controlling the ingress of moisture, ionic and gaseous species into the concrete. Once they get into concrete structure, the structure no longer maintains its structural integrity; the lifespan is reduced, and the general safety of the public is severely in danger. Many conventional techniques like application of chemical admixtures (plasticizers, water reducing agents etc.) are known which improve the workability of concrete by reducing intergranular friction finally affecting porosity. However, they come along with various disadvantages: a) incompatibility of protective layer and underlying layer due to differences in thermal expansion coefficient b) disintegration of protective layer over time c) need for constant maintenance along with contributing to pollution (Camaiti et al., 1988; Rodriguez- Navarro et al., 2003).

Due to these shortcomings, effect of microbial concrete on permeation properties was studied by different researchers. Permeability can be investigated by carbonation tests as it is increasingly apparent that decrease in gas permeability due to surface treatments results in an increased resistance towards carbonation and chloride ingress. Carbonation is related to the nature and connectivity of the pores, with larger pores giving rise to higher carbonation depths.

The biodeposition by microbial concrete should be regarded as a coating system. This is because of the fact that precipitation is mainly on the surface due to limited penetration of bacteria in the porous matrix. Ramakrishnan et al (1998) reported an increase in resistance of concrete towards alkali, freeze thaw attack, drying shrinkage and reduction in permeability upon application of bacterial cells.

De Muynck et al (2008b) studied the effect of biodeposition of calcite on permeability characteristics of mortar by *B. sphaericus*. The presence of biomass contributed to a large extent in the overall decrease of the gas permeability. Significant differences in carbonation depth between treated and untreated specimens were noticeable after 2 weeks of accelerated carbonation in treated mortar specimens. Bacterial treated specimens were found to have better resistance towards chloride penetration as compared to untreated mortar specimens. The increased resistance towards the migration of chlorides of cubes treated with biodeposition was similar to that of the acrylic coating and the water repellent silanes and silicones and larger than in the case of the silanes/siloxanes mixture, which were all reported to be effective in decreasing the rate of reinforcement corrosion (Basheer et al., 1997; Ibrahim et al., 1997).

Achal et al (2011a) reported the decreased water permeability of bioremediated cement mortar cubes treated by *Sporosarcina pasteurii* . The lower permeability of the bioremediated cubes compared with that of the control cubes was probably due to a denser interfacial zone formed because of calcite precipitation between the aggregate and the concrete matrix. The penetration of water at the sides was found to be higher than that at the top. This is due to better compaction and closing of pores at the top. This demonstrated the profound effect of microbial calcite on the permeability of concrete. The same group studied the effect of *Bacillus pasteurii* on water impermeability in concrete cubes and found the reduction in penetration of water which was more significant on the top side as compared to sides because of better compaction and closing of pores at the top surface (Achal et al., 2010b). Six times reduction in absorption of water was reported upon treatment of mortar cubes with *Bacillus* sp. CT-5 as compared to untreated specimens (Achal et al., 2011b).

3.1.3 Reduction in corrosion of reinforced concrete

Corrosion of steel and rebar structures in concrete is one of the main reasons for failure of structures. Corrosion initiates due to ingress of moisture, chloride ions and carbon dioxide through the concrete to the steel surface. Corrosion and permeation are somehow correlated. The permeability of water and pollutants are amongst the major threats to reinforced concrete. Such penetrations lead to ingress of moisture and chlorides which is responsible for early leakages and corrosion of steel. Corrosion products (iron oxides and hydroxides) lead to stresses that crack and spall the concrete cover which in turn exposes reinforcement to direct environmental attack that results in accelerated deterioration of the structure (Neville, 1995).

Application of microbial calcite may help in sealing the paths of ingress and improve the life of reinforced concrete structures (Jonkers et al., 2007; Fig. 3). Mukherjee et al (2010) reported four fold reduction in corrosion of reinforced concrete specimens upon application of *Bacillus sp. CT-5*. The same group observed reduction in water and chloride ion permeability upon use of calcite by *Sporosarcina pasteurii*. Qian et al (2010) used *B. pasteurii* to check its effect on permeability resistance and acid attack and reported that bacterial calcite improves surface permeability resistance and resist the attack of acid (pH> 1.5).

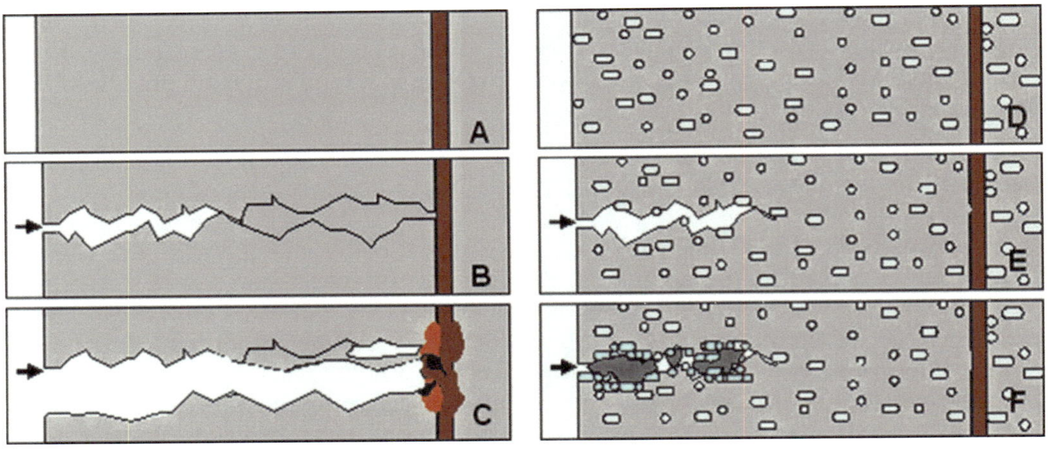

Fig. 3. Schematic drawing of conventional concrete (A–C) versus bacteria-based self-healing concrete (D–F). Crack ingress chemicals degrade the material matrix and accelerate corrosion of the reinforcement (A–C). Incorporated bacteria-based healing agent activated by ingress water seals and prevents further cracking (D–F) (Jonkers et al., 2007).

3.2 Microbial concrete in crack remediation

Use of microbial concrete has exhibited high potential for remediation of cracks in various structural formations such as concrete and granite (Gollapudi et al., 1995; Stocks-Fisher et al., 1999). Microbiologically enhanced crack remediation has been reported by Bang and Ramakrishnan (2001) where *Bacillus pasteurii* was used to induce calcium carbonate precipitation. Ramachandran et al (2001) proposed microbiologically enhanced crack remediation (MECR) in concrete. Specimens were filled with bacteria, nutrients and sand. Significant increase in compressive strength and stiffness values as compared to those without cells was demonstrated. The presence of calcite was limited to the surface areas of crack because bacterial cells grow more actively in the presence of oxygen. Extremely high pH of concrete germinated the need for providing protection to microbes from adverse environmental conditions. Polyurethanes were used as vehicle for immobilization of calcifying enzymes and whole cells because of its mechanically strong and biochemically inert nature (Klein & Kluge 1981; Wang & Ruchenstein, 1993). Bang et al (2001) investigated the encapsulation of bacterial cells in polyurethanes and reported positive potential of microbiologically enhanced crack remediation by polyurethane immobilized bacterial cells. They also studied the effect of immobilized bacterial cells on strength of concrete cubes by varying the concentration of immobilized cells per crack. Highest compressive strength was obtained with cubes remediated with 5 X 10^9 immobilized cells crack^{-1} for 7 days while after that, increase in strength was found to be marginal. SEM pictures also depicted the clear involvement of immobilized bacterial cells in sealing of cracks (Fig. 4).

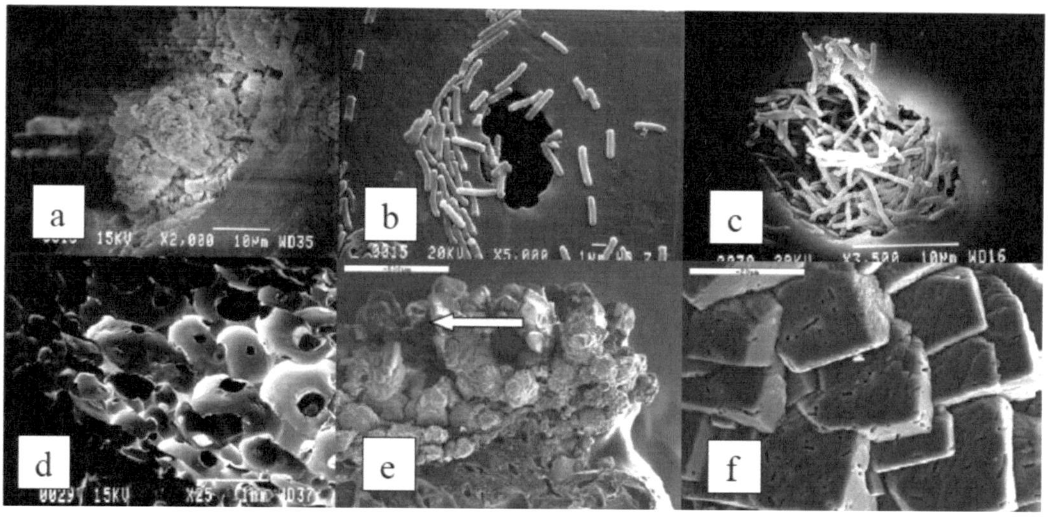

Fig. 4. Scanning electron micrographs of calcite precipitation induced by *B. pasteurii* immobilized in Polyurethanes (PU) (a) Porous PU matrix without microbial cells showing open-cell structures. Bar, 1 mm (b) Distribution of microorganisms on the PU surface. Bar, 1 µm (c) Microorganisms densely packed in a pore of the PU matrix. Bar, 10 µm. (d) Calcite crystals grown in the pore (shown in c) of the PU matrix. Bar, 10 µm. (e) Calcite crystals grown extensively over the PU polymer. Bar, 500 µm. (f) Magnified section pointed with an arrow in e shows crystals embedded with microorganisms. Bar, 20 µm. (Bang et al., 2001).

De Belie & De Muynck (2008) reported positive potential of microbiologically induced carbonate precipitation for the repair of cracks in concrete by *B. sphaericus* while Qian et al (2010) also reported that compressive strength of treated specimens could be restored to 84% upon treatment of bacterial calcite.

3.3 Microbial concrete in restoration of stone buildings

The pyramids of Egypt have been built with durable carbonate stones. With time, the calcareous matrix of the stone shows progressive increase in its porosity and a significant decrease of its mechanical characteristics due to the calcite leaching process (Amoroso, 1983). This leads to breakage of the materials into smaller particles and finally back into constituent minerals. Several conservative treatments are available with inorganic and organic products (Lazzarini & Laurenzi Tabasso, 1986) which can slow down the deterioration process of monuments. However, they offer several drawbacks due to their chemical composition and thermal expansion coefficient as they differ a lot from that of the stone. There is long-term incompatibility of the substrate and the new cement used for consolidation (Clifton, 1980) and the plugging of pores in the treated material induced by the new cement or protective layers (Lazzarini & Tabasso, 1980). These products are also formulated and applied in solvents at very low concentration that lead to huge waste of organic solvents in the environment. More over, their efficiency is inconsistent and in certain cases, they can have a detrimental effect for the conservation of the stone material itself. Shortcomings of conventional techniques have drawn the attention to bacterially induced carbonate precipitation for reducing the permeation properties and thereby, enhancing the durability properties of ornamental stone. Various researchers have applied this technology for remediation of stone (Le Me´tayer et al., 1999; Rodriguez-Navarro et al., 2003; Dick et al., 2006; Tiano et al., 1995, 1999, 2006; Jimenez- Lopez et al., 2007). Overview of different applications of microbial concrete on stone is given in table 2 (De Muynck et al., 2010).

Mediator	Organism/ molecule	Reference
Microorganisms	*Bacillus cereus*	Le Metayer- Levrel et al (1999)
	Micrococcus sp. Bacillus subtilis	Tiano et al (1999)
	Myxococcus xanthus	Rodriguez- Navarro et al (2003)
	Bacillus sphaericus	Dick et al (2006)
	Pseudomonas putida	May (2005)
Organic matrix molecules	*Mytilus californianus shell extracts* Aspartic acid Bacillus cell fragments	Tiano (1995) and Tiano et al (2006)
Activator medium	Microbiota inhabiting the stone	Jimenez- Lopez et al (2007)

Table 2. Different applications of microbial concrete on stone are given in table 2 (De Muynck et al., 2010).

3.3.1 Reduction in permeation of stone by microbial concrete

The main target for consolidation of stone aims at reducing the permeation and water absorption by providing surface treatment as this layer plays the most important protective role. Adolphe et al (1990) were among the first to consider the use of microbially induced carbonate precipitation (MICP) for the protection of ornamental stone. In addition, the role of calcite layer in providing resistance against erosion was reported. Bacterially induced calcium carbonate was compatible with the substrate and significantly reduced the water sorption of the treated stone (Le Me´ tayer-Levrel et al., 1999). However, the layer of the new cement induced by *B. cereus* was very thin - only a few microns-thick. Along with this, the formation of endospores and uncontrolled biofilm by *Bacillus* provides a potential drawback in stone conservation.

Rodriguez-Navarro et al (2003) proposed the use of *Myxococcus xanthus,* an abundant Gram-negative, non-pathogenic aerobic soil bacterium which belongs to a peculiar microbial group whose complex life cycle involves a remarkable process of morphogenesis and differentiation. This bacterium has been known to induce the precipitation of carbonates, phosphates and sulfates in a wide range of solid and liquid media (González-Mu˜noz et al., 1993, 1996; Ben Omar et al., 1995, 1998; Ben Chekroun et al., 2004 & Rodriguez-Navarro et al., 2007). Upon application of this bacterial suspension on stone specimens, no fruiting bodies were observed and there was no uncontrolled bacterial growth. Calcium carbonate cementation was observed up to a depth of several hundred micrometers (> 500μm) without any plugging or blocking of the pores. Plugging occurs mainly due to film formation by extracellular polymeric substance (EPS) (Tiano et al., 1999). In this case, only limited EPS production occurred in stones.

Tiano et al. (1999) commented on the use of viable cells for the formation of new minerals inside monumental stones. They studied the effect of microbial calcite crystals on Pietra di Lecce bioclastic limestone by using of *Micrococcus* sp., and *Bacillus subtilis*. Significant decrease in water absorption was observed due to the physical obstruction of pores. Furthermore, the authors commented on some possible negative consequences, such as (1) the presence of products of new formation, due to the chemical reactions between the stone minerals and some by-products originating from the metabolism of viable heterotrophic bacteria and (2) the formation of stained patches, due to the growth of air-borne micro-fungi related to the presence of organic nutrients necessary for bacterial development. In order to avoid these short comings, the authors proposed the use of natural and synthetic polypeptides to control the growth of calcite crystals in the pores. The use of organic matrix macromolecules (OMM) extracted from *Mytilus californianus* shells was proposed (Tiano et al., 1992; 1995) to induce the precipitation of calcium carbonate within the pores of the stone. In this case, there was slight decrease in porosity and water absorption by capillarity (Tiano, 1995). Due to the complexity of extraction procedure as well as low yields of usable product, this method was not beneficial (Tiano et al., 1999). In place of these bio inducing macromolecules (BIM) rich in aspartic acid groups, Tiano et al (2006) put forward the proposal of using acid functionalized proteins such as polyaspartic acid. For calcite crystal growth, calcium and carbonate ions were supplied by addition of ammonium carbonate and calcium chloride solution or a saturated solution of bicarbonate, supplemented in some cases by calcite nanoparticles so as to maintain saturated carbonate solution in the pores over a prolonged period. Spraying was used to introduce proteins, calcium ions and nanoparticles. This was found to be suitable for marble statues and objects of high aesthetic value which require minimum change in the chemistry of the object. The consolidating effect in this case was very low compared to ethylsilicates (Tiano et al., 2006).

Dick et al. (2006) reported 50% reduction in water absorption by treating limestone cubes with two strains of *B. sphaericus*. Various researchers in Biobrush consortium worked on improving the methodologies for delivering bacterial cells to stone surfaces and controllung side effects of bacteria to the stone. Various carrier materials were looked upon. Ranalli et al (1997) used sepiolite for delivering *Desulfovibrio vulgaris* and *D. desulfuricans* as it provided anaerobic conditions, humidity and shortened treatment time. Capitelli et al (2006) reported the superiority of Carbogel as a delivery system for bacteria because of higher retention of viable bacteria and less time needed for entrapment of cells by this. Zamarreno et al (2009) investigated the application of calcite crystals precipitated by fresh water bacteria on limestone and found significant reduction in pore sizes of the bacterial treated substrate stone specimens as compared to the untreated ones (Fig. 5). Bacterial calcite crystals had deposited around and inside open pore spaces. The application of calcite crystals resulted in filling 43-49% of the open pore spaces which was 20% more than the application of the medium alone.

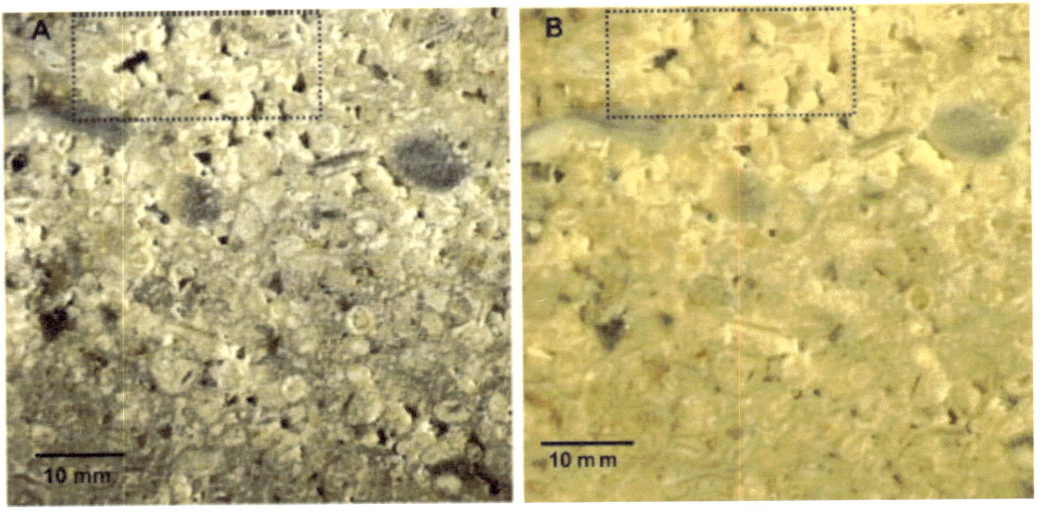

Fig. 5. Bacterial isolate on limestone before treatment (A) and (B) after Bacterial biocalcifying treatment for 21 days of incubation at 30°C (Zamarreno et al., 2009).

3.4 Bio-mediated ground improvement

The liquefaction of loose sands in hydraulic fills and manmade or natural (underwater) slopes is a major problem in geo-engineering. Piping and liquefaction are associated with sudden and catastrophic failures and often lead to loss of life and massive financial consequences. Mitigation strategies such as various methods of compaction or ground improvement methods such as jet-grouting or soil mixing are not always suitable or require high amounts of energy, high costs and materials with significant impact on the environment (van Paassen et al. 2010). Microbial-induced carbonate precipitation by urea hydrolysis has shown promising role for ground improvement. Recent research initiatives (Whiffin et al. 2005; Mitchell & Santamarina, 2005; DeJong et al. 2006; Whiffin et al. 2007; Ivanov & Chu 2008) have shown that the calcite crystals form cohesive "bridges" between existing sand grains, increasing strength and stiffness of sand with limited decrease in permeability. Proper understanding of key parameters, which control the *in situ* distribution of $CaCO_3$ and related engineering properties in both naturally and induced cemented sands, still prove to be insufficient and therefore represents the greatest challenge for further development of the bio-mediated ground stabilization technology.

In order to induce MICP in the soil subsurface, reagents and catalysts need to be injected and transported to the location where strengthening is required. Mixing bacteria and reagents prior to injection results in immediate flocculation of bacteria and crystal growth. While this method can be applied for treatment of surfaces, very coarse grained materials and mixed in place applications (Le Metayer- Levrel et al., 1999), this would cause rapid clogging of the injection well and surrounding pore space for many (fine) sands. In order to prevent crystal accumulation around the injection point and encourage a more homogeneous distribution of $CaCO_3$ over large distance, a two-phase injection for bacterial retainment has been suggested by Whiffin et al. (2007). Microbial transport in saturated porous media is well studied (Murphy & Ginn, 2000). Many physical, chemical and biological factors which influence the transport of bacteria have been investigated, including: fluid properties like chemistry and flow regime (Torkzaban et al., 2008), cell wall characteristics like hydrophobicity, charge and appendages (van Loosdrecht et al., 1987; Gilbert et al., 1991) and solid properties, like grain size distribution, surface texture and mineralogy (Scholl et al., 1990; Foppen and Schijven, 2005).

The fixation methodology, suggested by Whiffin et al. (2007), is based on the effect of ionic strength on microbial transport. To achieve homogeneous strength, Harkes et al. (2010) used BioGrout, a microbially induced calcium carbonate to improve the strength of ground. They developed a procedure to enhance fixation and distribution of bacterial cells and their enzyme activity in sand in order to improve the potential of microbially induced carbonate precipitation as ground reinforcement technique in fine-grained sand. The procedure comprises a multi-step injection in which first a bacterial suspension is introduced, potentially followed by a fixation fluid (i.e. a solution with high salt content) before the cementation fluid is introduced.

The new ground reinforcement techniques developed based on microbially induced carbonate precipitation (Whiffin et al. 2007; Harkes et al. 2010) use microbially catalyzed hydrolysis of urea to produce carbonate. In the presence of dissolved calcium this process leads to precipitation of calcium carbonate crystals, which form bridges between the sand grains and hence increase strength and stiffness. In addition to urea hydrolysis, there are many other microbial processes which can lead to the precipitation of calcium carbonate. van Paassen et al. (2010) evaluated various factors such as substrate solubility, $CaCO_3$ yield, reaction rate and type and amount of side-product. They found that the most suitable candidate as alternative MICP method for sand consolidation turned out to be microbial denitrification. In this process organic compounds, like acetate, can be oxidized to produce carbonate ions and alkalinity, which are required for the precipitation of calcite, while nitrate is reduced to nitrogen gas. Using calcium salts of both the electron donor and acceptor results in a high $CaCO_3$ yield. The rate of calcium carbonate formation by denitrification is far lower than the urease process, it requires further optimization for practical applications.

From the aforementioned applications, microbial concrete brings a new aspect to the construction industry. Promising results on the use of microorganisms for improvement of the durability of building materials have drawn the attention of research groups from all over the world. However, there are several challenges which must be addressed before wide acceptance of these strategies for construction materials.

4. Challenges to the study

4.1 Reduction of cost

One of the major factors hindering the use of MICCP technology is the high cost required for its production. The cost required is attributed to price of microbial product and the number of applications required. For better precipitation of carbonates, more time is required during which the building material has to be wet. With increasing times of precipitation, increasing amounts of EPS production, biofilm formation and hence, plugging can be expected. In order to ensure the presence of a sufficient amount of water, multiple applications of nutrients over several days (Le Metayer-Levrel et al., 1999) or the application of a carrier material (May, 2005) have been proposed. Both these measures add to the total cost of the treatment. De Muynck et al (2010) analyzed the cost of biodeposition treatment based on the price of the microorganisms and the price of the nutrients. The calculated price of 1 kg lyophilized bacteria was about US $1,500 (1,100 €) and 2–3 g m^{-2} is applied which costs about US $4 (3 €) m^{-2}. The cost of nutrients is estimated to be about US $250 (180 €) per kg. The dosage for biodeposition on concrete surface generally ranges between 0.04 and 0.08 kg m^{-2}, bringing the cost of nutrients to US $7–15 (5–10 €) m^{-2}. In case of carrier materials, the costs are even higher. Successful adoption and commercialization of this technique requires economical alternatives of the bacteria and the nutrients.

Economical alternatives to the medium ingredients, which can cost high as 60% of the total operating costs, need to be developed (Kristiansen, 2001). The nutritional profile of bacterial cultures indicate a high preference for protein based media as for *S. pasteurii* (Morsdorf & Kaltwasser, 1989). So for economizing this technology, researchers have looked for available cheap nutrient sources . There are many industrial effluents that are rich in proteins. If released in the altered form, they are hazardous for the atmosphere. So, the dual benefits of cost reduction and environment protection is feasible. Two such wastes are lactose mother liquor (LML) and corn steep liquor (CSL). Lactose mother liquor is an industrial effluent of the dairy industry. Its composition is given in table 3. Achal et al (2009a) investigated the effect of LML as sole source of growing bacterium *S. pasteurii* and compared the calcification effect of its usage. LML served as a better nutrient source for the growth of bacteria and also for calcite precipitation as compared to nutrient broth and yeast extract media which are quite expensive. Another by-product of the corn wet milling industry used by Achal et al (2010b) for economization of microbial calcite technology was corn steep liquor. Its constituents are listed in table 3. Corn steep liquor can typically be available locally with a price of nearly US $2 (1.5 €) per liter, which is very economical compared with standard nutrient medium. The biodeposition cost by this comes to US $0.5–1.0 (0.3–0.7 €) m^{-2}. 1.5 % CSL media along with NaCl, urea and $CaCl_2$ was used to investigate its effect on water and chloride ion permeability along with compressive strength improvement in cement mortar cubes and compared with Nutrient broth and yeast extract. The performance of CSL was significantly better than standard laboratory nutrients in terms of microbial concrete production. CSL offered an economic advantage over the standard nutrient medium and the overall process cost reduced dramatically. The usage of such byproducts not only reduces the cost, but also serve to prevent environmental pollution.

4.2 Usage of industrial by products

The traditional construction materials such as concrete, bricks, hollow blocks, solid blocks, pavement blocks and tiles are being produced from the existing natural resources. This is damaging the environment due to continuous exploration and depletion of natural resources. Many authorities and investigators are now working to have the privilege of reusing the wastes in environmentally and economically sustainable ways (Aubert et al., 2006). Different types of wastes along with their recycling and utilization potentials are listed in table 4.

LML		CSL	
Component	**Measure**	**Component**	**Measure**
pH	6.2	pH	3.86
Solids (%)	5.5	Solids (%)	46-50
Lactose (%)	15.4	Carbohydrates (%)	5.8
Proteins (%)	8	Proteins (%)	24
Fats (%)	2	Fats (%)	1
Ash (%)	0.53	Minerals (%)	8.8
Calcium (mg/l)	353	Arginine (%)	0.4
Phosphorus (mg/l)	35	Cystine (%)	0.5
Potassium (mg/l)	186	Glycine (%)	1.1
Sodium (mg/l)	44	Isoleucine (%)	0.9
Chloride (mg/l)	90	Inositol (mg/ 100g)	602
Sulphur (mg/l)	15	Choline (mg/ 100g)	351

Table 3. Physico chemical characteristics of lactose mother liquor (LML) and corn steep liquor (CSL) (Achal et al., 2009a, 2010b).

Fly ash (FA) generated during the combustion of coal for energy production is one of the industrial byproduct that is recognized as an environmental pollutant. Addition of fly ash to concrete has become a common practice in recent years. Reports have been published concerning the effect of fly ash on concrete porosity and resistivity, pore solution chemistry, oxygen and chloride ion diffusivity, carbonation rates and passivation (Mangat & Gurusamy, 1987; Thomas & Matthews, 1992; Montemor et al., 2000). Rice husk ash (RHA) obtained from burning of rice husk is another major agricultural byproduct and can be used successfully in construction materials such as bricks and blocks without any degradation in the quality of products (Nasly & Yassin., 2009). The utilization of above mentioned by-products as partial replacement of clay in bricks can serve important economical, environmental and technical benefits such as the reduced amount of waste materials, cleaner environment, reduced energy requirement, durable service performance during service life and cost effectiveness. However, problems associated with ash bricks are low strength, higher water adsorption, low resistance to abrasion, low fire resistance and high porosity (Kumar & Palit, 1994). An attempt has been made to study the role of microbial calcite to enhance the durability of ash bricks (FA and RHA) and it is found to be very effective in reducing permeability and decreasing water absorption leading to enhanced durability of ash bricks (Dhami et al. 2011).

Similar efforts need to be accomplished for other industrial wastes as well as with more detailed studies of energy inputs for alternative materials, so that efficiency can be improved in comparison to traditional materials.

Type of wastes	Source details	Recycling and utilization potentials
Industrial waste (inorganic)	Coal combustion residues, fly ash, steel slag, construction debris	Bricks, blocks, tiles, cement, paint, fine and coarse aggregates, concrete, wood substitute products, ceramic products
Agro waste (organic)	Baggage, rice and wheat straw and husk, saw mill waste, ground nut shell, jute, sisal, cotton stalk	Cement boards, particle boards, insulation boards, wall panels, roof sheets, binder, fibrous building panels, bricks, acid proof cement, coir fiber, reinforced composites, polymer composites
Mining/ mineral wastes	Coal washeries waste, mining waste tailing from iron, copper, zinc, gold industries	Bricks, fine and coarse lightweight aggregates, tiles
Non hazardous waste	Waste gypsum, lime sludge, lime stone waste, broken glass and ceramics	Blocks, bricks, cement clinker, hydraulic binder, fibrous gypsum boards, gypsum plaster, super sulfated cement
Hazardous waste	Contaminated blasting materials, galvanizing waste, metallurgical residues, sludge from waste water and waste water treatment plants	Boards, bricks, cement, ceramics, tiles

Table 4. Different types and sources of solid wastes and their recycling and utilization potentials for construction materials (adapted from Pappu et al., 2007).

4.3 Concentration of ammonia and salts

The production of ammonia during hydrolysis of urea might raise some issues of environmental concern because of the fact that atmospheric ammonia is being recognized as a pollutant. Atmospheric ammonia is known to contribute to several environmental problems, including direct toxic effects on vegetation, atmospheric nitrogen deposition, leading to the eutrophication and acidification of sensitive ecosystems, and to the formation of secondary particulate matter in the atmosphere, with effects on human health, atmospheric visibility and global radiative balance (Sutton et al., 2008).

However, when the concentration of ammonia generating compounds does not exceed the concentration of the calcium salt, it is possible to decrease the emission of ammonia to a great extent. The presence of ammonium might also present some risks to the stone. First of all, the presence of an ammonium salt might present some risks related to salt damage. Secondly, ammonium can be converted to nitric acid by the activity of nitrifying bacteria, resulting in damage to the stone. So, if higher concentrations of ammonium are to be produced, as might be the case for the hydrolysis of urea, the use of a paste might offer an attractive solution. The latter is one of the most commonly applied methods for the removal of salts from building materials (Woolfitt & Abrey, 2008; Carretero et al., 2006).

After ammonia, calcium salt concentration has a major impact on the performance of treatment. High dosage of calcium salt could also lead to an accumulation of salts in the stone, which could give rise to efflorescence or damage related to crystallization. So, detailed studies need to be done in this regard for prevention of such harmful effects.

4.4 Survival of bacteria

The size of bacterial inoculum and survival of bacteria potentially influences bacterial calcification. Zamarreno et al (2009) studied the survival of bacteria inside carbonate crystals for up to 330 days. Significant reduction in the viable cells was noticed after 13 days interval and after 330 days, no cells are viable.

Survival of bacteria inside cracks and other building materials also needs to be studied in detail, such that the efficacy of this treatment can be evaluated.

4.5 Microbial concrete in low carbon buildings

The art and science of building constructions started with the usage of natural materials like soils, stones, leaves, unprocessed timber etc. Hardly any energy is spent in manufacturing and usage of such materials, but they are not much in use because of durability issues unlike materials like burnt clay bricks, lime, cast iron products, aluminium, steel, Portland cement, etc. These modern materials require huge energy reservoirs, are non-recyclable, and as well as are harmful to the environment. The construction sector is responsible for major input of energy resulting in large share of CO_2 emissions (22% in India) into the atmosphere (Reddy & Jagadish.,2001). The emission of these green house gases during manufacturing processes of building materials is contributing a lot to global warming. Its time to put emphasis on reducing the emission of these gases into the atmosphere and save energy by minimizing usage of conventional building materials, methods, techniques and working on some other substitutes. For reduction of indirect energy use in building materials, either alternative for bricks, steel and cement have to be found, or vigorous energy conservation measures in these segments of industry have to be initiated. Energy requirements for production and processing of different building materials, CO_2 emissions and the implications on environment have been studied by many researchers (Suzki *et al.*, 1995; Oka *et al.*, 1995; Debnath *et al.*, 1995; table 5).

Type of material	Thermal energy (MJ/kg)
Cement	5.85
Lime	5.63
Lime Pozollana	2.33
Steel	42.00
Aluminium	236.80
Glass	25.80

Table 5. Energy in basic building materials (Reddy & Jagadish, 2001).

Reddy & Jagadish (2001) reported soil cement blocks with 6-8% cement content to be most energy efficient building material. These materials have low cost, are easily recyclable and environmental friendly as the soils are mixed with additives like cement, limestone etc. As there is no burning involved, this type of stabilized mud block helps in conserving huge amounts in energy. Attempts are being made in our lab to apply Microbial calcite technology to these eco friendly low carbon building materials so that it can pave the way for more sustainable, cheap and durable building materials.

4.6 Enhancing the efficiency of calcifying bacteria

It is imperative to use local bacterial strains that are well conditioned for the environment for production of microbial calcite. There are evidences that indicate the implementation of mutagenesis through UV for strain improvement (Wu et al., 2006). Attempts to enhance the efficiency of calcifying bacterial culture of *S. pasteurii* were done by Achal et al (2009a). They got more promising results with UV induced mutants of as compared to wild type strains for consolidation of sand columns. Further steps are required to produce better calcifying cultures through other methods so as to get better consolidation of various construction materials.

4.7 Other factors

Rodriguez-Navarro et al (2003) reported that fast precipitation of bacterial carbonates could result in a lower efficiency of the calcite deposition. Along with this, the presence of well developed rhombohedral calcite crystals result in a more pronounced consolidating effect compared to the presence of tiny acicular vaterite crystals. So, detailed studies need to focus on different types of nutrients and metabolic products used for growing calcifying microorganisms, as they influence survival, growth, biofilm and crystal formation. More work should be done on the retention of nutrients and metabolic products in the building material. Detailed microbial ecology studies are also needed in order to ascertain the effects of the introduction of new bacteria into the natural microbial communities, the development of the communities at short, mid and long-term, and the eventual secondary colonization of heterotrophic microorganisms using bacterial organic matter and dead cells, such as actinomycetes, fungi, etc. Until now, the practical applications of microbial calcite technology have been mainly limited to France where it has been applied on several historic monuments including a part of the Notre Dame de Paris (De Muynck et al. 2010).

5. Conclusion

Microbial concrete technology has proved to be better than many conventional technologies because of its eco- friendly nature, self healing abilities and increase in durability of various building materials. Work of various researchers has improved our understanding on the possibilities and limitations of biotechnological applications on building materials. Enhancement of compressive strength, reduction in permeability, water absorption, reinforced corrosion have been seen in various cementitious and stone materials. Cementation by this method is very easy and convenient for usage. This will soon provide the basis for high quality structures that will be cost effective and environmentally safe but, more work is required to improve the feasibility of this technology from both an economical and practical viewpoints.

6. References

Achal, V.; Mukherjee, A. & Reddy, M.S. (2011a). Effect of calcifying bacteria on permeation properties of concrete structures, *J Ind Microbiol Biotechnol,* Vol. 38, pp. 1229-1234.

Achal, V.; Mukherjee, A. & Reddy, M.S. (2010). Biocalcification by *Sporosarcina pasteurii* using Corn steep liquor as nutrient source, *J. Ind. Biotechnol.,* Vol. 6, pp.170-174.

Achal, V.; Mukherjee, A. & Reddy, M.S. (2011b). Microbial Concrete: A Way to Enhance the Durability of Building Structures, *J Mater Civ Eng,* Vol. 23, pp. 730-734.

Achal, V.; Mukherjee, A.; Basu, P.C. & Reddy, M.S. (2009b). Strain improvement of *Sprosarcina pasteurii* for enhanced urease and calcite production, *J Ind Microbiol Biotechnol,* Vol. 36, pp. 981-988.

Achal, V.; Mukherjee, A.; Basu, P.C. & Reddy, M.S. (2009a). Lactose mother liquor as an alternative nutrient source for microbial concrete production by *Sporosarcina pasteurii, J Ind Microbiol Biotechnol,* Vol.36, pp. 433-438.

Adolphe, J.P.; Loubière, J.F.; Paradas, J. & Soleilhavoup, F. (1990). Procédé de traitement biologique d'une surface artificielle, European patent 90400G97.0. (after French patent 8903517, 1989).

Amoroso, G. & Fassina, V. (1983). Stone Decay and Conservation, *Elsevier*, Amsterdam.

Aubert, J.E.; Husson, B. & Sarramone, N. (2006). Utilization of Municipal Solid Waste Incineration (MSWI) Fly Ash in Blended Cement: Part 1: Processing and Characterization of MSWI Fly Ash, *J. Hazardous Mater.,* Vol.136, pp. 624-631.

Bachmeier, K.L.; Williams, A.E.; Warmington, J.R. & Bang, S.S. (2002). Urease activity in microbiologically-induced calcite precipitation, *J. Biotechnol.,* Vol. 93, pp.171- 181.

Banfield, J.F. & Hamers, R.J. (1997). Processes at minerals and surfaces with relevance to microorganisms and prebiotic synthesis, pp. 81-117. In Banfield, J.F. & Nealson, K.H. (eds.), *Geomicrobiology: interactions between Microbes and Minerals.* Mineralogical Society of America, Washington D.C.

Banfield, J.F. & Zhang, H. (2001). Nanoparticles in the environment, *Rev Mineral Geochem,* Vol. 44, pp.1-58.

Bang, S.S.; Galinat, J.K. & Ramakrishnan, V. (2001). Calcite precipitation induced by polyurethane-immobilized *Sporosarcina pasteurii, Enzyme Microb. Technol.,* Vol.28, pp.404-409.

Basheer, L. & Cleland, D.J. (2006). Freeze–thaw resistance of concretes treated with pore liners, *Construction and Building Materials,* Vol. 20, pp. 990–998.

Basheer, P.A.M.; Basheer, L.; Cleland, D.J. & Long, A.E. (1997). Surface treatments for concrete: assessment methods and reported performance, *Construction and Building Materials* Vol.11, pp. 413–429.

Bäuerlein, E. (2003). Biomineralization of unicellular organisms: An unusual membrane biochemistry for the production of inorganic nano- and microstructures, *Angew Chem Int,* Vol. 42, pp. 614-641.

Ben Chekroun, K.; Rodriguez-Navarro, C.; González-Munoz, M.T.; Arias, J.M.; Cultrone, G.& Rodriguez-Gallego, M. (2004). Precipitation and growth morphology of calcium carbonate induced by *Myxococcus xanthus*: implications for recognition of bacterial carbonates, *J. Sediment. Res.,* Vol. 74, pp. 868–876.

Ben Omar, N.; González-Mu˜noz, M.T. & Pe˜nalver, J.M.A. (1998). Struvite crystallization on *Myxococcus* cells, *Chemosphere,* Vol.36, pp. 475–481.

Ben Omar, N.; Martínez-Ca˜namero, M.; González-Mu˜noz, M.T.; Maria Arias, J. & Huertas, F. (1995). *Myxococcus xanthus* killed cells as inducers of struvite crystallization Its possible role in the biomineralization processes, *Chemosphere,* Vol. 30, pp. 2387–2396.

Beveridge, T.J. (1989). Role of cellular design in bacterial metal accumulation and mineralization, *Annu Rev Microbiol,* Vol. 43, pp.147-171.

Boquet, E.; Boronat, A. & Ramos-Cormenzana, A. (1973). Production of calcite (calcium carbonate) crystals by soil bacteria is a general phenomenon, *Nature,* Vol.246, pp. 527–529.

Braissant, O.; Verrecchia, E. & Aragno, M. (2002). Is the contribution of bacteria to terrestrial carbon budget greatly underestimated?, *Naturwissenschaften,* Vol. 89, pp.366–370.

Camaiti, M.; Borselli, G. & Matteol, U. (1988). Prodotti consalidanti impiegati nelle operazioni di restauro, *Edilizia,* Vol.10, pp. 125–134.

Cappitelli, F.; Zanardini, E.; Ranalli, G.; Mello, E.; Daffonchio, D. & Sorlini, C. (2006). Improved methodology for bioremoval of black crusts on historical stone artworks by use of sulfate-reducing bacteria, *Appl. Environ. Microbiol.,* Vol. 72, pp. 3733-3737.

Carretero, M.I.; Bernabé, J.M.& Galan, E. (2006). Application of sepiolite-cellulose pastes for the removal of salts from building stones, *Appl. Clay Sci.,* Vol. 33, pp. 43–51.

Castanier, S.; Le Metayer-Levrel, G. & Perthuisot, J.P. (1999). Ca-carbonates precipitation and limestone genesis—the microbiogeologist point of view, *Sediment. Geol.,* Vol. 126, pp. 9–23.

Chafetz, H.S. & Buczynski, C. (1992). Bacterally induced lithification of microbial mats. *Palaios,* Vol. 7, pp. 277–293.

Chaturvedi, S.; Chandra, R. & Rai, V. (2006). Isolation and characterization of *Phragmites australis* (L.) rhizosphere bacteria from contaminated site for bioremediation of colored distillery effluent, *Ecol. Eng.,* Vol. 27, pp. 202–207.

Claisse, P. A.; Elsayad, H. A. & Shaaban I. G. (1997). Absoprtion and sorptivity of cover concrete, *Journal of Materials in Civil Engineering,* Vol.9, pp.105-110.

Clifton, J.R. & Frohnsdorff, G.J.C. (1982). Stone consolidating materials: a status report. In: *Conservation of Historic Stone Buildings and Monuments,* National Academy Press,Washington, DC, pp. 287–311.

Clifton, J.R. (1980). Stone consolidating material: a status report. In: Conservation of Historic Stone Buildings and Monuments. Department of Commerce, National Bureau of Standards, Washington D.C.

De Belie, N. & De Muynck,W. (2008). Crack repair in concrete using biodeposition. *In: Proc. of ICCRR,* Cape Town, South Africa.

De Muynck, W.; Belie, N. & Verstraete, W. (2010). Microbial carbonate precipitation in construction materials: a review, *Ecol Eng,* Vol.36, pp.118–136.

De Muynck, W.; Cox, K.; De Belie, N. & Verstraete, W. (2006). Bacterial carbonate precipitation as an alternative surface treatment for concrete. *Construction and Building Materials,* doi:10.1016/ j.conbuildmat.2006.12.011.

De Muynck,W.; Cox, K.; De Belie, N. & Verstraete,W. (2008a). Bacterial carbonate precipitation as an alternative surface treatment for concrete, *Constr. Build. Mater.,* Vol. 22, pp. 875–885.

De Muynck,W.; Debrouwer, D.; De Belie, N. & Verstraete,W. (2008b). Bacterial carbonate precipitation improves the durability of cementitious materials, *Cem. Concr. Res.*, Vol. 38, pp.1005–1014.

Debnath, A.; Singh, S.V. & Singh, Y.P. (1995). Comparative assessment of energy requirements for different types of residential buildings in India, *Energy and Buildings*, Vol. 23, pp. 141–146.

DeJong, J.T., Fritzges, M.B., Nusslein, K., 2006. Microbially induced cementation to control sand response to undrained shear. *J. Geotech. Geoenviron.*, Vol.132, pp.1381–1392.

Delgado Rodrigues, J. (2001). Consolidation of decayed stones. A delicate problem with few practical solutions. *In: Lourec¸ o, P.B., Roca, P. (Eds.)*, International Seminar on Historical Constructions. Guimarães, Portugal.

Dhami, N.K; Mukherjee, A. and Reddy, M.S. (2011). Bacterial calcite as sealant of fly ash mortar substrates. In: 31st Cement and Concrete Science Conference on Novel Developments and Innovation in Cementitious Materials, Imperial College London, U.K (12-13 September 2011).

Dick, J.; De Windt, W.; De Graef, B.; Saveyn, H.; van der Meeren, P.; De Belie, N. & Verstraete, W. (2006). Bio-deposition of a calcium carbonate layer on degraded limestone by *Bacillus* species, *Biodegradation* Vol.17, pp.357–367.

Douglas, S. & Beveridge, T.J. (1998). Mineral formation by bacteria in natural microbial communities, *FEMS Microbiol Ecol*, Vol. 26, pp. 79–88.

Ehrlich, H.L. (1996). How microbes influence mineral growth and dissolution, *Chem. Geol.*, Vol. 132, pp.5–9.

Ehrlich, H.L. (2002). *Geomicrobiology*, 4th ed. New York: Marcel Dekker.

Ferris, F.G. & Stehmeier, L.G. (1992). Bacteriogenic mineral plugging. USA Patent US5143155.

Folk, R. (1993). SEM imaging of bacteria and nanobacteria in carbonate sediments and rocks, *J Sedim Petrol*, Vol. 63, pp. 990–999.

Foppen, J.W.A. & Schijven, J.F. (2005). Transport of *E. coli* in columns of geochemically heterogeneous sediment. *Water Res.* Vol. 39, pp. 3082–3088.

Fortin, D.; Ferris, F.G. & Beveridge, T.J. (1997). Surface-mediated mineral development by bacteria, *Rev Mineral*, Vol.35, pp. 161-180.

Fujita, Y.; Redden, G.D.; Ingram, J.C.; Cortez, M.M.; Ferris, F.G. & Smith, R.W. (2004). Strontium incorporation into calcite generated by bacterial ureolysis, *Geochim. Cosmochim. Acta*, Vol. 68 , pp. 3261–3270.

Ghosh, P.; Mandal, S.; Chattopadhyay, B.D. & Pal, S. (2005). Use of microorganism to improve the strength of cement mortar, *Cem. Concr. Res.*, Vol. 35, pp.1980–1983.

Gilbert, P.; Evans, D.J.; Evans, E.; Duguid, I.G. & Brown, M.R.W. (1991). Surface characteristics and adhesion of *Escherichia coli* and *Staphylococcus epidermidis*. *J. Appl. Bacteriol.* Vol. 71, pp. 72–77.

Gollapudi, U.K.; Knutson, C.L.; Bang, S.S. & Islam, M.R. (1995). A new method for controlling leaching through permeable channels, *Chemosphere*, Vol. 30, pp. 695–705.

González-Mu˜noz, M. ; Arias, J.M.; Montoya, E. & Rodriguez-Gallego, M. (1993). Struvite production by *Myxococcus coralloides D.*, *Chemosphere*, Vol. 26, pp. 1881– 1887.

González-Mu˜noz, M.T.; Omar, N.B. ; Martínez-Ca˜namero, M.; Rodríguez-Gallego, M.; Galindo, A.L. & Arias, J. (1996). Struvite and calcite crystallization induced by cellular membranes of *Myxococcus xanthus.*, *J. Cryst. Growth,* Vol. 163, pp. 434–439.

Hamilton, W.A. (2003). Microbially influenced corrosion as a model system for the study of metal microbe interactions: a unifying electron transfer hypothesis, *Biofoulin,* Vol. 19, pp. 65–76.

Hammes, F. & Verstraete,W. (2002). Key roles of pH and calcium metabolism in microbial carbonate precipitation, *Rev. Environ. Sci. Biotechnol.* Vol. 1, pp. 3–7.

Hammes, F.; Seka, A.; de Knijf, S. & Verstraete, W. (2003). A novel approach to calcium removal from calcium-rich industrial wastewater, *Water Res.*, Vol. 37, pp. 699–704.

Harkes, M.P.; van Paassen, L.A.; Booster, J.L.; Whiffin, V.S. & van Loosdrechta, M.C.M. (2010). Fixation and distribution of bacterial activity in sand to induce carbonate precipitation for ground reinforcement *Ecol. Eng.* Vol, 36, pp. 112–117

Ibrahim, M.; Al-Gahtani, A.S.; Maslehuddin, M. & Almusallam, A.A. (1997). Effectiveness of concrete surface treatment materials in reducing chloride-induced reinforcement corrosion, *Construction and Building Materials,* Vol. 11, pp. 443–451.

Ivanov, V. & Chu, J. (2008). Applications of microorganisms to geotechnical engineering for bioclogging and biocementation of soil in situ. *Rev. Environ. Sci. Biotechnol.* Vol. 7, pp. 139-153.

Jimenez-Lopez, C.; Rodriguez-Navarro, C.; Pi ˜nar, G.; Carrillo-Rosúa, F.J.; Rodriguez-Gallego, M. & González-Mu˜noz, M.T. (2007). Consolidation of degraded ornamental porous limestone by calcium carbonate precipitation induced by the microbiota inhabiting the stone, *Chemosphere,* Vol.68, pp.1929–1936.

Jonkers, H. (2007). Self healing concrete: a biological approach. *In: van der Zwaag, S. (Ed.), Self Healing Materials: An alternative Approach to 20 Centuries of Materials Science,* Springer, Netherlands, pp. 195–204.

Jonkers, H.M. & Schlangen, E. (2007). Crack repair by concrete-immobilized bacteria.In: Schmets, A.J.M., van der Zwaag, S. (Eds.), *Proc. of First International Conference on Self Healing Materials,* Noordwijk, The Netherlands.

Jonkers, H.M.; Thijssen, A.; Muyzer, G.; Copuroglu, O. & Schlangen, E. (2010). Application of bacteria as self-healing agent for the development of sustainable concrete, *Ecol Eng,* Vol.36, pp.230-235.

Keefe, W.E. (1976). Formation of crystalline deposits by several genera of the family *Enterobacteriaceae, Infect Immun,* Vol.14, pp.590–592.

Khan, M. I. (2003). Isoresponses for strength, permeability and porosity of high performance mortar, *Build. Environ.,* Vol.38, pp.1051–1056.

Klein, J. & Kluge, M. (1981). Immobilization of microbial cells in polyurethane matrices, *Biotechnol. Lett.* Vol. 3, pp. 65-70.

Knorre, H. & Krumbein, K.E. (2000). Bacterial calcification. In: Riding, E.E., Awramik, S.M. (Eds.), *Microbial Sediments.* Springer–Verlag, Berlin, pp. 25–31.

Konhauser, K. O.(1998). Diversity of bacterial iron mineralization, *Earth-Sci Rev,* Vol. 43, pp. 91-121.

Kristiansen, B. (2001). Process Economics. In C. Ratledge and B. Kristiansen (Ed.), Biotechnology, 2nd edn. Cambridge University Press, Cambridge.

Krumbein, W.E. (1979). Photolithotrophic and chemoorganotrophic activity of bacteria and algae as related to beachrock formation and degradation (Gulf of Aqaba, Sinai), *Geomicrobiol J*, Vol.1, pp.139–203.

Kumar, D. & Palit, A. (1994). *Proceedings of American Power Conference*, Chicago pp.471-476.

Lazzarini, L.& Laurenzi Tabasso, M. (1986). Il Restauro della Pietra. CEDAM, Padova.

Lazzarini, L., Tabasso, M., 1980. Il Restauro della Pietra. CEDAM, Padova

Le Metayer-Levrel, G.; Castanier, S.; Orial, G.; Loubiere, J.F. & Perthuisot, J.P. (1999). Applications of bacterial carbonatogenesis to the protection and regeneration of limestones in buildings and historic patrimony. *Sedimentary Geology* Vol. 126, pp. 25–34.

Le Metayer-Levrel, G.; Castanier, S.; Orial, G.; Loubiere, J.F. & Perthuisot, J.P. (1999). Applications of bacterial carbonatogenesis to the protection and regeneration of limestones in buildings and historic patrimony, *Sediment. Geol.*, Vol. 126, pp. 25–34.

Lowenstam, H.A. & Weiner, S. (1989). *On Biomineralization.* Oxford University Press, New York.

Lowenstam, H.A. (1981). Minerals formed by organisms, *Science,* Vol. 211, pp.1126-131.

Mangat, P.S. & Gurusamy, K. (1987). Chloride diffusion in steel fibre reinforced concrete containing pfa, *Cem. Concr. Res.*, Vol.17, pp. 640–650.

May, E. (2005). Biobrush research monograph: novel approaches to conserve our European heritage. EVK4-CT-2001-00055.

Mc Connaughey, T.A. & Whelan, J.F. (1997). Calcification generates protons for nutrient and bicarbonate uptake, *Earth Sci. Rev.*, Vol. 42, pp. 95–117.

McKay, D.S., Gibson, E.K., Thomas-Keprta, K.L., Vali, H., Romanek, C.S., Clement, S., Chiller, X.D.F, Maechling, C.R. & Zare, R.N. (1996). Search for past life on Mars: Possible relic biogenic activity in Martian meteorite ALH84001, *Science,*Vol. 273, pp. 924–930.

Mitchell, J.K. & Santamarina, J.C. (2005). Biological considerations in geotechnical engineering. *J. Geotech. Geoenviron. Eng.* Vol. 131, pp. 1222-1233.

Mörsdorf, G. & Kaltwasser, H. (1989). Ammonium assimilation in *Proteus vulgaris, Bacillus pasteurii, and Sporosarcina ureae, Arch. Microbiol.,* Vol.152, pp. 125–131.

Monger, H.C.; Daugherty, L.A.; Lindemann, W.C. & Liddell, C.M.(1991). Microbial precipitation of pedogenic calcite, *Geology,* Vol. 19, pp. 997–1000.

Montemor, M.F.; Simoes, A.M.P. & Salta, M.M. (2000). Effect of fly ash on concrete reinforcement corrosion studied by EIS, *Cem. Concr. Comp.*, Vol. 22, pp. 175- 185.

Moon, H.Y.; Shin, D.G. & Choi, D.S. (2007). Evaluation of the durability of mortar and concrete applied with inorganic coating material and surface treatment system, *Construction and Building Materials,* Vol. 21, pp. 24–33.

Moropoulou, A.; Kouloumbi, N.; Haralampopoulos, G.; Konstanti, A. & Michailidis, P. (2003). Criteria and methodology for the evaluation of conservation interventions on treated porous stone susceptible to salt decay, *Prog. Org. Coat.,* Vol. 48, pp. 259–270.

Mukherjee, A.; Achal, V. & Reddy, M.S. (2010a). In Search of a Sustainable Binder in Building Materials, *Annals of Ind. Acad. Of Engg.,* Vol. VII, pp. 41-51.

Murphy, E.M. & Ginn, T.R. (2000). Modeling microbial processes in porous media. *Hydrogeol. J.* Vol. 8, pp.142–158.

Nasly, M.A. & Yassin, A.A.M. (2009). Sustainable Housing Using an Innovative Interlocking Block Building System. *In Proceedings of the Fifth National Conference on Civil Engineering (AWAM '09): Towards Sustainable Development,* Kuala Lumpur, Malaysia, 130-138.

Nemati, M. & Voordouw, G. (2003). Modification of porous media permeability, using calcium carbonate produced enzymatically in situ, *Enzyme Microb. Technol.*, Vol. 33, pp. 635–642.

Nemati, M.; Greene, E.A. & Voordouw, G. (2005). Permeability profile modification using bacterially formed calcium carbonate: comparison with enzymic option, *Process Biochem,* Vol. 40, pp. 925–933.

Neville, A.M. (1995). Properties of concrete, 4th ed. Essex, England, UK: Longman Group Limited

Nolan, E.; Basheer, P.A.M. & Long, A.E. (1995). Effects of three durability enhancing products on some physical properties of near surface concrete, *Construction and Building Materials,* Vol. 9, pp. 267–272.

Oka, T.; Suzuki, M. & Konnya, T. (1993). The estimation of energy consumption and amount of pollutants due to the construction of buildings, *Energy and Buildings,* Vol.19, pp. 303–311.

Pappu, A.; Saxena, M. & Asolekar, S.R. (2007). Solid Wastes Generation in India and their Recycling Potential in Building Materials, *Building and Environment,* Vol.42, pp. 2311-2320.

Park, Sung-Jin.; Park, Yu- Mi.; Chun, Woo-Young.; Kim, Wha-Jung & Ghim, Sa-Youl.(2010). Calcite-Forming Bacteria for Compressive Strength Improvement in Mortar, *J. Microbiol. Biotechnol.*, Vol. 20, pp. 782–788.

Pentecost, A. & Bauld, J. (1988). Nucleation of calcite on the sheaths of cyanobacteria using a simple diffusion cell, *Geomicrobiol J,* Vol.6, pp. 129-135.

Qian, C.; Wang, R.; Cheng, L. & Wang, J. (2010). Theory of microbial carbonate precipitation and its application in restoration of cement-based materials defects, *Chin J Chem,* Vol. 28, pp. 847–857.

Ramachandran, S.K.; Ramakrishnan, V. & Bang, S.S. (2001). Remediation of concrete using micro-organisms, *ACI Materials journal,* Vol. 98, pp. 3–9.

Ramakrishnan, S.K.; Panchalan, R.K. & Bang, S.S. (2001). Improvement of concrete durability by bacterial mineral precipitation, *Proceedings of 11th International Conference on Fracture,* Turin, Italy, 2001.

Ramakrishnan, V. (2007). Performance characteristics of bacterial concrete—a smart biomaterial. *In: Proceedings of the First International Conference on Recent Advances in Concrete Technology,*Washington, DC, 2007, pp. 67–78.

Ramakrishnan, V.; Bang, S. S. & Deo, K. S. (1998). A novel technique for repairing cracks in high performance concrete using bacteria, *Proc. Int. Conf. on High Performance High Strength Concrete,* Perth, Australia, pp. 597–618.

Ranalli, G.; Chiavarini, M.; Guidetti, V.; Marsala, F.; Matteini, M.; Zanardini, E. & Sorlini, C. (1997). The use of micro-organisms for the removal of sulphates on artistic stoneworks, *Int. Biodeterior. Biodegrad.,* Vol. 40, pp.255–261.

Reddy, B.V.V. & Jagadish, K.S. (2003). Embodied energy of common and alternative building materials and technologies, *Energy and Buildings,* Vol. 35, pp.129–137.

Rivadeneyra, M. A. G.; Delgado, A.; Ramos-Cormenzana & Delgado, R. (1998). Biomineralization of carbonates by *Halomonas eurihalina* in solid and liquid media with different salinities: crystal formation sequence, *Res. Microbiol.*, Vol. 149, pp. 277–287.

Rivadeneyra, M.A.; Parraga, J.; Delgado, R., Ramos-Cormenzana, A. & Delgado, G. (2004). Biomineralization of carbonates by *Halobacillus trueperi* in solid and liquidmedia with different salinities, FEMS Microbiol. Ecol., Vol. 48, pp. 39–46.

Rodriguez-Navarro, C.; Jimenez-Lopez, C.; Rodriguez-Navarro, A.; González-Mu˜noz, M.T. & Rodriguez-Gallego, M. (2007). Bacterially mediated mineralization of vaterite, *Geochim. Cosmochim. Acta*, Vol. 71, pp. 1197–1213.

Rodriguez-Navarro, C.; Rodriguez-Gallego, M.; Ben Chekroun, K. & Gonzalez-Munoz, M.T. (2003). Conservation of ornamental stone by *Myxococcus xanthus* induced carbonate biomineralization, *Appl Env Microbiol*, Vol. 69, pp. 2182–2193.

Sarda, D.; Choonia, H.S.; Sarode, D.D. & Lele, S.S. (2009). Biocalcification by *Bacillus pasteurii* urease: a novel application, *J Ind Microbiol Biotechnol*, Vol. 36, pp.1111-1115.

Scholl, M.A.; Mills, A.L.; Herman, J.S. & Hornberger, G.M. (1990). The influence of mineralogy and solution chemistry on the attachment of bacteria to representative aquifer materials. *J. Cont. Hydrol.* Vol. 6, pp. 321–336.

Simon, M.A.; Bonner, J.S.; Page, C.A.; Townsend, R.T.; Mueller, D.C.; Fuller, C.B. & Autenrieth, R.L. (2004). Evaluation of two commercial bioaugmentation products for enhanced removal of petroleum from a wetland, *Ecol. Eng.*, Vol. 22, pp. 263–277.

Stocks-Fischer, S.; Galinat, J.K. & Bang, S.S. (1999). Microbiological precipitation of $CaCO_3$, *Soil Biol. Biochem.*, Vol.31, pp. 1563–1571.

Stumm,W. & Morgan, J.J. (1981). Aquatic Chemistry, 2nd edition. John Wiley, NewYork.

Sutton, M.; Reis, S. & Baker, S. (2008). Atmospheric ammonia: detecting emission changes and environmental impact. *In: Results of an Expert Workshop Under the Convention on Long-Range Transboundary Air Pollution*, Springer, pp. 490.

Suzuki, M.; Oka, T. & Okada, K. (1995). The estimation of energy consumption and CO_2 emission due to housing construction in Japan, *Energy and Buildings*, Vol. 22, pp. 165–169.

Thomas, M.D.A. & Matthews, J.C. (1992). The permeability of fly ash concrete, *Mater. Struct.*, Vol. 25, pp. 388–396.

Thomas-Keprta, K.L.; McKay, D.S.; Wentworth, S.J.; Stevens, T.O.; Taunton, A.E.; Allen, C.; Coleman, A.; Gibson, E.K. & Romanek, C.S. (1998). Bacterial mineralization patterns in basaltic aquifers: Implications for possible life in Martian meteorite ALH84001, *Geology*, Vol. 26, pp.1031–1034.

Tiano, P. (1995). Stone reinforcement by calcite crystal precipitation induced by organic matrix macromolecules, *Stud. Conserv.*, Vol. 40, pp.171–176.

Tiano, P.; Biagiotti, L. & Mastromei, G. (1999). Bacterial bio-mediated calcite precipitation for monumental stones conservation: methods of evaluation, *J. Microbiol. Methods*, Vol. 36, pp.139–145.

Tiano, P.; Addadi, L. & Weiner, S. (1992). Stone reinforcement by induction of calcite crystals using organic matrix macromolecules: feasibility study. *In: 7th International Congress on Deterioration and Conservation of Stone*, Lisbon, pp. 1317–1326.

Tiano, P.; Cantisani, E.; Sutherland, I. & Paget, J.M. (2006). Biomediated reinforcement of weathered calcareous stones, *J. Cult. Herit.*, Vol. 7, pp. 49–55.

Torkzaban, S.; Tazehkand, S.S.; Walker, S.L.; Bradford, S.A. (2008). Transport and fate of bacteria in porous media: coupled effects of chemical conditions and pore space geometry. *Water Resources Res*. Vol. 44, pp. 1–12.

van Loosdrecht, M.C.M.; Lyklema, J.; Norde,W.; Schraa, G. & Zehnder, A.J.B. (1987). The role of bacterial cell wall hydrophobicity in adhesion. *Applied and Environmental Microbiology* Vol. 53, pp. 1893–1897.

van Paassen, L.A.; Daza, C.M.; Staal, M.; Sorokin, D.Y.; van der Zon, W. & van Loosdrecht, M.C.M. (2010). Potential soil reinforcement by biological denitrification. *Ecol. Eng.* Vol. 36, pp. 168–175.

Vasconcelos, C.; McKenzie, J.A.; Bernaconi, S.; Grujic, D. & Tien, A.J. (1995). Microbial mediation as a possible mechanism for natural dolomite formation at low temperatures, *Nature,* Vol. 377, pp. 220–222.

Wang, X. & Ruchenstein, E. (1993). Preparation of porous polyurethane particles and their use of enzyme immobilization, *Biotechnol. Prog.* Vol. 9, pp. 661-665.

Warren, L.A.; Maurice, P.A.; Parmar, N. & Ferris, F.G. (2001). Microbially mediated calcium carbonate precipitation: implications for interpreting calcite precipitation and for solid-phase capture of inorganic contaminants, *Geomicrobiol. J.,* Vol. 18, pp. 93–125.

Whiffin, V.S.; Lambert, J.W.M.: & Van Ree, C.C.D. (2005). "Biogrout and Biosealing –Pore - Space Engineering with Bacteria." Geostrata - Geo Institute for ASCE 5 (5). Vol. 13-16, pp. 36.

Whiffin, V.S.; van Paassen, L. & Harkes, M.P. (2007). Microbial carbonate precipitation as a soil improvement technique, *Geomicrobiol. J,* Vol. 24, pp. 417–423.

Woolfitt, C. & Abrey, G. (2008). Poultices: the true or plain poultice and the cleaning and desalination of historic masonry and sculpture. Retrieved august 2008, from http://www.buildingconservation.com/articles/poultices/poultice.htm.

Worrell, E.; Price, L.; Martin, N.; Hendriks, C. & Ozawa Meida, L. (2001). Carbon dioxide emissions from the global cement industry, *Annu. Rev. Energy Environ.,* Vol. 26, pp.303–329.

Wu, C.; Teo, V.S.; Farrell, R.L.; Bergquist, P.L. & Nevalainen, K.M. (2006). Improvement of the secretion of extracellular proteins and isolation and characterization of the amylase I (amy1) gene from *Ophiostoma floccosum*, *Gene*, Vol.384, pp. 96-103.

Zamarreno, D.V.; Inkpen, R. and May, E. (2009). Carbonate crystals precipitated by freshwater bacteria and their use as a limestone consolidant, *Applied and Environment Microbiology,* Vol. 75, pp. 5981-5990.